Quantum Business Dynamics™

A quantum leap in business transformation

Textbook Edition

J. Oliver Glasgow

with contributions from Alan Nekhom and Stuart Davis

ISBN:

979-8-218-31030-1

LICENSE INFORMATION

This material is protected by copyright law. It represents a proprietary and copyrighted methodology developed by Data Stewardship Solutions Inc.

Unauthorized use, copying, or dissemination of this material, including the unauthorized use of the brand concepts, "Quantum Business Dynamics™" and "QBD™", is strictly prohibited without prior written approval from Data Stewardship Solutions Inc. Any unauthorized use may result in legal action.

The contents of this material can be further licensed for training, certification, and implementation purposes. For inquiries regarding licensing use, certification in training, management, or leadership of QBD™, and detailed information on available licensing options and certification programs, please contact Data Stewardship Solutions at info@datastewardship.solutions

Please note that the purchase of this book does not grant an automatic license. Licensing is a separate process that requires explicit approval and agreement from Data Stewardship Solutions Inc. By licensing and implementing this methodology, organizations gain access to the knowledge, tools, and support necessary to effectively apply the QBD™ principles within their operations.

Revised first edition

Dedication

To our loving spouses and families, who have provided us with unwavering support, encouragement, and understanding throughout the process of bringing this book to fruition. Your love, patience, and sacrifice are the bedrock of our success, and we dedicate this book to you with heartfelt gratitude.

To our team of dedicated professionals, who have tirelessly worked with us to bring QBD into the fore as a game-changing methodology. Your hard work, commitment, and passion for excellence have made this journey possible, and we are forever grateful for your contributions.

Thank you to our clients around the world for being a part of this journey with us. Your efforts and dedication have been instrumental in shaping the success of QBD, and we couldn't have done it without you. This book is dedicated to you as a testament to our shared vision and commitment to excellence.

Contents

Acknowledgments

We would like to express our deepest appreciation to our colleagues, mentors, and advisors who have generously shared their expertise, insights, and feedback throughout the research and writing process. Your guidance and support have been invaluable and greatly contributed to the development of this manuscript.

We would also like to acknowledge our editors, proofreaders, and designers who have worked tirelessly to refine the manuscript and bring it to its final form. Your attention to detail and dedication to excellence have truly made this a polished and professional product.

Forward

Welcome to the future of business with Quantum Business Dynamics™ (QBD™), the game-changing methodology that's transforming the way businesses operate globally. Imagine understanding and utilizing the fundamental building blocks of business success with ease—the four essential elements known in QBD as particles: Source, Bridge, Target, and Binary Link. These particles combine to create the ultimate synergy that enables organizations to optimize operations, mitigate risks, and unlock new opportunities, all with ease and simplicity.

Quantum Business Dynamics™ (QBD™) is a holistic, data-driven framework that enables businesses to optimize decision-making, streamline processes, and enhance collaboration by leveraging advanced technologies such as distributed ledger technology, artificial intelligence, and machine learning analytics.

According to a recent study by McKinsey & Company, companies that embrace digital transformation through innovative methodologies, like QBD, can achieve up to 15% revenue growth and a 20% reduction in costs (McKinsey 2023). This is a testament to the power of digital transformation, and with QBD, your business can unlock its full potential.

The best part? QBD is easy to learn and apply. By simply identifying and classifying all particles in your business, executives can gain new levers of control, shape the trajectory of their business, eliminate the biggest risks and avoid missed opportunities. Whether you're a CEO or an entry-level employee, QBD provides a methodology that can be learned quickly and requires very little training to implement.

To make it even easier for businesses to apply QBD, we will present easy-to-adopt interfaces for humans and machines to help capture the data about the four particles without requiring clients to purchase any new platforms or machines. These standard interfaces and agreements support data-center-to-data-center connectivity in many of the commonly used clouds today, making it easy for businesses to integrate QBD seamlessly into their existing infrastructure.

In addition to its simplicity and ease of use, QBD also provides built-in immutable ethics and established rules for Bridges, including privacy rights and rights equivalent to ownership, leading to a better, more trusted approach to business. With such standards in place, businesses can achieve something previously thought to be impossible: a standard for Duty of Care.

As a busy executive, you may be wondering why you would want to learn yet another methodology and adopt more tech approaches. Well, let's take a moment to think about the three things that keep executives up at night: optimizing operations, mitigating risks, and unlocking new opportunities. The three most crucial responsibilities that executives need to address will be highlighted next up. If these crucial tasks have piqued your interest, then read on!

THE THREE MOST IMPORTANT TASKS

1. Optimizing Operations

First and foremost, optimizing operations is the foundation of any successful business. Whether you are a small startup or a large corporation, the importance of efficient, adaptable, and agile operations cannot be overstated. In today's fast-paced and competitive world, achieving operational excellence is paramount.

For example, in the manufacturing sector, managing the supply chain is often a major challenge. Deloitte, in a 2020 report, emphasized the need for manufacturers to adopt an end-to-end digital supply network (DSN) to improve visibility, agility, and efficiency (Deloitte 2020). By applying the principles of QBD and strategically leveraging its particles, businesses can address these challenges head-on, leading to a more streamlined and effective supply chain.

But optimizing operations is not just about managing the supply chain. It also includes optimizing other critical functions of a business, such as finance, human resources, and marketing. For instance, QBD's particles can be used to streamline the financial processes of a business by automating the validation of invoices, purchase orders, and payments. This leads to fewer errors, faster processing times, and a more streamlined financial management system overall.

Similarly, by leveraging QBD's particles, human resources departments can easily manage employee data, recruitment processes, and benefits administration. The application of QBD's principles ensures that human resources teams are able to efficiently manage the day-to-day functions of the department, allowing them to focus on the more strategic initiatives that will drive the business forward.

And let's not forget marketing. By applying QBD's particles, businesses can optimize their marketing efforts by tracking customer data and analyzing trends to better understand their audience. This leads to more effective marketing campaigns, increased customer engagement, and increased revenue.

But the benefits of QBD don't stop there. In fact, QBD's principles are so powerful that they can be applied across all industries and sectors, from healthcare to finance to retail. For example, in the healthcare industry, QBD's particles can be used to optimize patient care by streamlining the patient intake process, automating appointment reminders, and even predicting patient outcomes based on data analysis.

And in the finance industry, QBD's particles can be used to mitigate fraud by automating validation processes and ensuring that only validated Sources are bridged to Targets. This reduces the risk of financial losses due to fraudulent activity, ultimately leading to a more secure financial system.

One anecdote that illustrates the power of QBD's principles involves a small retail business struggling to keep up with the demands of their customers. By implementing QBD's particles, they were able to streamline their supply chain, automate their financial processes, and optimize their marketing efforts. As a result, they were better able to meet the needs of their customers, resulting in increased sales and revenue growth.

Optimizing operations is the first step towards achieving business success. By leveraging the power of QBD's particles, businesses can streamline their operations, reduce costs, and increase productivity. And with QBD's principles being applicable across all industries and sectors, there's no limit to the benefits that can be achieved.

2. Mitigating Risks

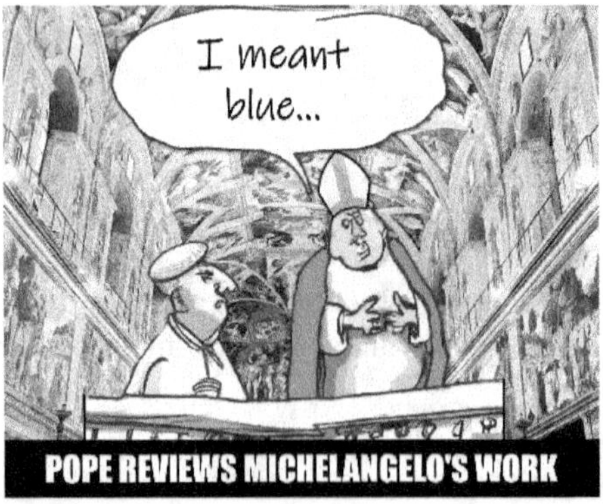

No business is immune to risk, but managing risks is essential for ongoing success. By leveraging the power of the particles, executives can identify and mitigate risks across a wide range of areas, including financial, operational, strategic, and compliance-related concerns.

In the financial services sector, regulatory compliance is a significant concern. A study by Accenture found that the cost of non-compliance for financial institutions is

2.71 times greater than the cost of maintaining or meeting compliance requirements (Accenture 2017). With QBD's methodology, organizations can identify and address such risks, ensuring that they remain compliant and avoid potential pitfalls.

But compliance is just one area of risk that businesses face. Operational risks such as supply chain disruptions, cyber-attacks, and natural disasters can also have a significant impact on businesses. By applying QBD's particles, businesses can identify potential risks and take proactive measures to prevent them from becoming larger problems down the line.

Another area of risk that businesses face is strategic risk, which involves making decisions that could negatively impact the long-term success of the company. By applying QBD's principles, businesses can use data analysis to make informed decisions, reducing the likelihood of making decisions that could result in a significant risk to the company.

And let's not forget about compliance-related risks. QBD's methodology provides a framework for identifying and addressing these risks, ensuring that organizations remain compliant with regulatory requirements and avoid potential legal issues.

There once was a small business owner who thought they had everything under control until a major supplier suddenly went bankrupt. The business owner had never considered the possibility of a supplier going out of business, and as a result, was left scrambling to find a new supplier at the last minute. By leveraging QBD's particles, businesses can proactively identify potential risks, including supplier bankruptcy, and take steps to mitigate those risks before they become larger problems.

Mitigating risks is crucial to the ongoing success of any business. By leveraging the power of QBD's particles, executives can identify potential risks and take proactive measures to prevent them from becoming larger problems down the line. Whether it's financial, operational, strategic, or compliance-related concerns, QBD's methodology provides a framework for addressing these risks and ensuring that organizations remain on the path to success.

3. Unlocking New Opportunities

Every executive dreams of discovering untapped opportunities for growth and innovation, but it can be challenging to identify and capitalize on them amidst the noise of daily business operations. QBD's particles provide the clarity needed to see beyond the horizon and find those elusive opportunities.

For instance, let's take the example of a small coffee shop owner who dreams of expanding their business but doesn't know where to start. By applying QBD's methodology, they can identify potential opportunities such as opening a second location, introducing a loyalty program, or adding new menu items that appeal to a wider audience. These opportunities might seem daunting at first, but with QBD's particles, the small business owner can break them down into smaller, more manageable parts and take steps towards making their dreams a reality.

Another example is in the fashion industry, where fast-changing trends and fickle consumer preferences can make it challenging for executives to identify and capitalize on new opportunities. However, with QBD's methodology, businesses can use data analysis to identify emerging trends, customer preferences, and market opportunities, allowing them to pivot quickly and stay ahead of the competition.

And let's not forget about the rapidly evolving digital landscape. By applying QBD's principles, businesses can harness the power of technology to unlock new opportunities for growth and innovation. For example, the use of AI and machine learning can help businesses better understand customer behavior and preferences, leading to more targeted marketing campaigns and improved customer experiences.

Optimizing operations, mitigating risks, and unlocking new opportunities is essential for business growth and innovation. By applying QBD's methodology and harnessing the power of its particles, businesses can identify and capitalize on new opportunities, pivot quickly in response to changing market conditions, and most importantly, bring to heal the four scariest risks while simultaneously stimulating the four most languishing missed opportunities.

To underscore the critical importance of not neglecting these three key domains, we can simply examine the four most significant risks and missed opportunities confronting today's businesses, as recounted by individuals like yourself. This peek under the rug may be uncomfortable, but it can enhance our comprehension and practical application of QBD to your enterprise.

THE FOUR BIGGEST RISKS

You have Risks

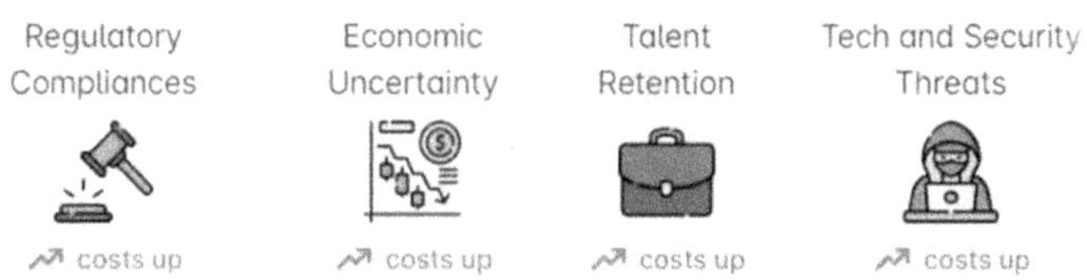

1. Regulatory Compliance

Compliance risks can pose a significant challenge to organizations across industries, ranging from financial institutions to healthcare providers. In recent years, companies have faced massive fines for failing to comply with regulations, making it essential for businesses to stay ahead of the curve.

One such example of the importance of compliance can be seen in the case of a biological data privacy fine. In 2020, a major pharmaceutical company was fined $5.1 million for failing to adequately protect sensitive patient data (CBS 2020). This violation of regulatory requirements not only resulted in a hefty fine but also negatively impacted the company's reputation, highlighting the importance of maintaining compliance in the healthcare industry.

Another example of compliance risks can be seen in the emerging medical cannabis industry. In 2021, a medical cannabis producer in Canada was hit with a hefty fine of $300,000 for violating regulations on selling unlicensed cannabis products (Global News 2021). This violation not only resulted in a significant financial penalty but also tarnished the company's reputation, demonstrating the importance of adhering to industry standards.

In the financial industry, compliance risks can lead to significant fines and reputational damage. In 2019, a major Australian bank was hit with a record-breaking fine of $530 million for noncompliance with open banking regulations (The Guardian 2020). This case highlights the importance of staying up to date with regulatory requirements and implementing internal policies to ensure compliance.

By leveraging QBD's principles, executives can identify and mitigate compliance risks before they become costly violations. By staying informed of changing regulations and adhering to industry standards, businesses can avoid fines, sanctions, and reputational damage, ultimately protecting their bottom line.

QBD's methodology provides a framework for identifying and addressing compliance risks across a range of industries, from healthcare to finance. By breaking down compliance risks into manageable parts and providing a clear understanding of regulatory requirements, QBD's particles can help businesses stay ahead of the curve and avoid costly violations.

Compliance risks can pose a significant challenge to businesses, ranging from reputational damage to massive fines. By adopting QBD's methodology and leveraging the power of its particles, businesses can identify and mitigate compliance risks before they become costly violations, ultimately protecting their bottom line and ensuring long-term success.

2. Economic Uncertainty

Economic uncertainty is a constant challenge for businesses across all industries. Unexpected events, market fluctuations, and global crises can cause financial risks that threaten a company's bottom line. By implementing QBD's methodology, businesses can proactively identify and mitigate these risks, protecting their financial well-being.

For example, the travel and tourism industry faced significant economic uncertainty during the COVID-19 pandemic, as governments worldwide implemented travel restrictions and lockdowns. Businesses in this sector that had not taken steps to mitigate financial risks faced devastating losses. On the other hand, if companies had implemented QBD's methodology, they would have been better prepared to weather the storm. A report by the World Tourism Organization highlighted the importance of such preparedness in the wake of economic uncertainty caused by crises (World Tourism Organization. 2020).

The retail industry is another sector that has been impacted by economic uncertainty. With the rise of e-commerce and changing consumer behavior, brick-and-mortar stores have faced increasing competition and financial risks. In addition, tariffs and trade disputes between countries have led to currency fluctuations, affecting global supply chains and causing additional financial risks for retailers. By using QBD's methodology to assess and mitigate these risks, retailers can remain competitive and financially stable.

Finally, the healthcare industry has also faced economic uncertainty in recent years. One example is the rising cost of pharmaceuticals, which has put pressure on healthcare providers to manage their budgets effectively. Failure to do so can result in reduced access to medication, compromising patient care. In addition, the industry has faced significant legal and regulatory changes, such as the Affordable Care Act in the United States, which has caused financial uncertainty for healthcare providers. By implementing QBD's methodology, healthcare providers can mitigate these risks and remain financially stable while continuing to provide quality care to their patients.

Economic uncertainty is a fact of life for businesses across all industries. However, by using QBD's methodology to assess and mitigate financial risks, organizations can protect their bottom line and remain competitive in a constantly changing economic landscape. By being prepared and proactive, businesses can weather the storm of economic uncertainty and come out stronger on the other side.

3. Talent Acquisition and Retention

Talent retention is one of the biggest challenges facing companies today. In a world where competition for top talent is fierce, businesses must take proactive measures to attract and retain the best employees. Failure to do so can result in significant financial costs, as well as a loss of intellectual capital and institutional knowledge.

One of the key issues facing companies in the area of talent retention is the ability to attract high-quality candidates. This is particularly true in industries with high demand for skilled labor, such as technology, healthcare, and finance. In these industries, the competition for top talent is fierce, and companies must work harder than ever to stand out from the crowd.

Another challenge facing companies is the retention of existing talent. High employee turnover rates can be incredibly costly for businesses, both in terms of financial costs and the loss of valuable institutional knowledge. Furthermore, high turnover rates can damage a company's reputation, making it more difficult to attract new talent in the future.

To address these challenges, companies can turn to QBD's methodology. By adopting our easy interfaces to your already rigorous hiring process that verifies previously validated required credentials and certifications, companies can ensure that they are attracting the best and brightest candidates, and just as importantly, that they can later prove regulatory compliance. This can help to establish a culture of excellence, where only the most qualified and capable individuals are hired.

In addition to hiring the right people, companies can also take steps to retain their existing talent. This includes offering a wide range of perks, benefits, and training opportunities, as well as opening up opportunities for cross-pollination, lateral promotions, and quality of life enhancements.

For example, by adopting QBD particle definitions for Sources and

Targets in the area of talent retention, your company can see an automated version of what one company had to achieve manually. Zappos, the online shoe and clothing retailer, offers its employees a wide range of perks and benefits, including free lunches, on-site gym facilities, and regular team-building activities. Additionally, the company encourages cross-pollination by allowing employees to switch departments and work on new projects. But even Zappos faces the issues of employees not fully realizing all the benefits they have coming, and that's just money falling on the floor for too many companies today.

By adopting QBD's principles and taking a proactive approach to talent retention, companies can capture and improve their already well thought-through approaches to attract and retain the best employees, while also reducing turnover rates and associated costs. This can help to establish a culture of excellence, where employees are valued and empowered to achieve their full potential.

4. Tech and Security Threats

In today's interconnected world, businesses face a multitude of tech and security threats that can disrupt operations and damage their reputation. From cyberattacks to equipment failure, these risks can be difficult to manage, particularly when addressed in a siloed approach. Companies must adopt a comprehensive strategy for mitigating these threats, and that's where QBD's methodology comes into play.

According to the consulting firm McKinsey, companies must take a holistic approach to address tech and security threats. This involves evaluating the organization's risk profile, developing a risk management framework, and establishing governance structures to oversee risk management (McKinsey & Company 2014). QBD's methodology provides a framework for identifying and addressing operational risks before they become major issues, helping organizations stay ahead of the curve.

One of the biggest challenges facing companies today is the lack of integration between various systems and tools. While each tool and process may be world-class on its own, without proper measurement and integration, it becomes difficult to identify the relationships and human agreements between them. This can create gaps in the security infrastructure that cyber attackers can exploit. The QBD methodology emphasizes the importance of measuring and monitoring all aspects of the organization's operational infrastructure, allowing businesses to identify potential vulnerabilities and address them before they become a threat.

Additionally, businesses must also address the human element of tech and security threats. In many cases, employees are the weakest link in the security chain, inadvertently or intentionally exposing the organization to threats. This can be addressed through employee training, awareness programs, and implementing robust access controls. The QBD methodology emphasizes the importance of cross-functional collaboration and communication, which can help identify and mitigate security risks originating from within the organization.

To illustrate the importance of addressing tech and security threats, let's consider a few examples from different industries. In the healthcare sector, a data breach can have catastrophic consequences, exposing sensitive patient information and damaging the organization's reputation. In 2019, a medical center in Texas was fined $2.3 million for violating HIPAA regulations related to data privacy and security (U.S. Department of Health and Human Services 2020). By implementing QBD's methodology, the organization could have identified and addressed potential vulnerabilities in its data privacy and security infrastructure, avoiding the hefty fine and reputational damage.

In the retail sector, companies face the threat of cyberattacks that can compromise customer data and lead to financial losses. In 2020, a major retail chain was hit by a ransomware attack that compromised customer data and cost the company millions of dollars in recovery efforts (CNBC 2020). By adopting QBD's methodology, the organization could have implemented a more robust security framework, with proper measurement and integration between various systems and tools, and potentially avoided the attack altogether.

Finally, in the banking sector, open banking regulations have opened up new opportunities for innovation and growth, but also introduced new security risks. Banks must ensure that customer data is protected and secure, and compliance with open banking regulations is critical. In 2019, a UK-based bank was fined £44 million for non-compliance with open banking regulations (Financial Conduct Authority). By adopting QBD's methodology, the organization could have identified potential compliance risks and implemented a framework for addressing them proactively.

Tech and security threats are a significant risk to organizations in today's interconnected world. By adopting a comprehensive approach to risk management and leveraging the power of QBD's methodology, businesses can identify and address operational risks before they become major issues, ensuring their long-term success and growth.

If we were to stop there, the story would indeed be pretty grim. However, the good news is that there is a way to offset the costs of unmitigated risks: by maximizing opportunities. But if your company is missing out on these opportunities, then you are in double jeopardy. Before we delve into the solution, let's take a moment to explore the four key missed opportunities that business leaders have been sharing with us. By addressing these missed opportunities and leveraging QBD's methodology, businesses can unlock significant growth and innovation potential while mitigating the risks that threaten their bottom line.

THE FOUR BIGGEST MISSED OPPORTUNITIES

1. Missed Innovation and Growth

Rapid advancements in technology present a wealth of opportunities for businesses across all sectors. However, many organizations fail to capitalize on these developments, either due to a lack of awareness or an inability to adapt quickly. One of the biggest culprits behind missed innovation and growth opportunities is a breakdown in communication within organizations.

According to a recent study by McKinsey, 86% of executives and employees cite a lack of collaboration or ineffective communication for workplace failures (McKinsey & Company 2012). This breakdown in communication can cause important data to remain siloed within departments, hindering a company's ability to innovate and grow.

In the energy sector, the role of Landsmen in securing a party's rights to a share in the profits from an energy project, such as an oil drilling, can be complicated. Landsmen play a critical role in the early stages of a project, helping to identify potential sites for drilling and negotiating contracts with landowners. However, failure to properly communicate and coordinate between different groups, including the Landsmen and the technical team, can lead to missed opportunities and costly mistakes.

Furthermore, the importance of accurate and up-to-date data cannot be overstated. In the energy sector, the success of a project is heavily reliant on geological data and other technical information. Failure to effectively communicate and share this data between different groups can lead to missed opportunities and increased operational risks. By applying QBD's methodology and harnessing the power of its particles, businesses in the energy sector can improve communication and collaboration between different groups, ensuring that all parties have access to the data they need to make informed decisions and capitalize on new opportunities.

Another industry affected by a lack of communication is the retail sector. A report by Retail Dive found that 70% of retail executives believe that a lack of collaboration between departments inhibits innovation and growth (Retail Dive 2018). The same report also found that 53% of retail executives believed that communication breakdowns lead to missed opportunities for business growth.

The consequences of a lack of communication and understanding of important data can be significant. A recent study by Harvard Business Review found that companies in the bottom quartile of communication effectiveness experience 57% more project failures (Harvard Business Review 2017). In contrast, companies in the top quartile of communication effectiveness achieve on average 25% higher gross margins and have a three times higher likelihood of outperforming their competitors.

By applying the principles of QBD, businesses can bridge the communication gap and unlock missed innovation and growth opportunities. QBD's methodology emphasizes collaboration and cross-functional teams, which can help break down departmental silos and stimulate communication and innovation. Additionally, QBD's particles can help businesses identify important data and ensure that it is communicated effectively throughout the organization.

One example of QBD's success in addressing communication breakdowns can be seen in the manufacturing industry. A report by Deloitte found that companies that implemented a cross-functional team approach to innovation achieved a 60% improvement in time-to-market and a 35% increase in product success rates (Deloitte 2018). The implementation of QBD methodology offers businesses the chance to foster collaboration and cross-functional teams, which can lead to enhanced innovation and growth prospects.

A lack of communication and understanding of important data can lead to missed innovation and growth opportunities for businesses across all sectors. By adopting QBD's methodology, businesses can break down departmental silos, nurture collaboration and cross-functional teams, and unlock missed opportunities for innovation and growth.

2. Missed Digital Transformation

Digital transformation is essential for businesses across all sectors to remain competitive and adapt to the ever-changing landscape. However, many organizations struggle to implement successful digital transformation strategies due to communication breakdowns and the lack of traceability of data between groups within the company.

One industry that has struggled with digital transformation is entertainment, particularly in the area of copyrights law. The rise of streaming services has created new challenges for content creators and distributors, leading to an increase in copyright infringement cases. In many instances, this could have been prevented with better communication and data traceability between departments, ensuring that all parties are aware of the proper rights and permissions for each piece of content.

Another industry that has faced digital transformation challenges is the finance sector, particularly in the area of crypto wallets. As the use of cryptocurrencies becomes more widespread, many financial institutions have struggled to adapt to the new technology. The lack of traceability of data and the inability to verify previously validated facts can lead to security breaches and financial losses. By implementing QBD's methodology and adding bridges between particles within existing systems, processes, and devices, financial institutions can mitigate these risks and ensure the security of their clients' assets.

Similarly, companies that raise capital for other companies, such as venture capital firms, also face challenges in digital transformation. The lack of traceability of data between departments can lead to missed investment opportunities and a lack of synergy between portfolio companies. By adopting QBD's principles and ensuring proper communication and data traceability, venture capital firms can improve their investment strategies and drive growth for their portfolio companies.

Digital transformation is crucial for businesses in all sectors to remain competitive and adapt to the changing landscape. However, communication breakdowns and the lack of traceability of data can impede the transformation process. By adopting QBD's methodology and adding bridges between particles within existing systems, processes, and devices, organizations can overcome these challenges and drive innovation and growth.

3. Missed Sustainability and Social Responsibility

Sustainability and social responsibility are not just buzzwords anymore. In today's world, they are essential aspects of any successful business. Unfortunately, many organizations fail to recognize the potential that sustainability and social responsibility offer for growth and innovation. As writer and entrepreneur Richard Branson once said, "Taking care of our planet and taking care of business are not mutually exclusive. For business to be successful over the long term, it must operate in an environment that is healthy, stable, and sustainable."

The missed opportunity for sustainability and social responsibility comes in many forms. It could be failing to reduce waste, ignoring social justice issues, or neglecting the impact of business operations on the environment. For example, a 2019 report by the Ellen MacArthur Foundation found that the fashion industry produces 10% of global carbon emissions, more than all international flights and maritime shipping combined (Ellen MacArthur Foundation 2017). This highlights the massive impact that businesses can have on the environment and the missed opportunity to reduce that impact.

The missed opportunity for sustainability and social responsibility also has a significant impact on consumers' buying habits. A recent survey by Nielsen found that 66% of consumers globally are willing to pay more for products and services that come from companies committed to positive social and environmental impact (Nielson 2018). This represents a significant opportunity for businesses to capture market share and drive growth by demonstrating a commitment to sustainability and social responsibility.

But the benefits of sustainability and social responsibility go beyond just customer preferences. They can also lead to increased employee engagement and retention. According to a study by Cone Communications, 64% of millennials consider a company's social and environmental commitments when deciding where to work (Cone Communications 2016). This represents a massive, missed opportunity for companies that fail to prioritize sustainability and social responsibility in their operations and culture.

To truly harness the potential of sustainability and social responsibility, businesses must adopt a holistic approach that integrates these principles into every aspect of their operations. This requires a cultural shift that prioritizes long-term sustainability over short-term profits. As sustainability advocate Paul Hawken once said, "Business is the only mechanism on the planet today powerful enough to produce the changes necessary to reverse global environmental and social degradation."

The good news is that the principles of QBD can help businesses take the necessary steps towards sustainability and social responsibility. By utilizing the particles to identify areas of improvement and establishing metrics to track progress, businesses can create a culture that prioritizes sustainability and social responsibility. This includes measures such as reducing waste and carbon emissions, ensuring ethical sourcing and production, and engaging with local communities.

4. Missed Operational Cost Savings

As businesses strive to maintain profitability and growth in a highly competitive landscape, finding ways to save on operational costs can make all the difference. However, many organizations fail to fully capitalize on this opportunity, either due to a lack of understanding or an inability to identify areas for improvement.

The missed opportunity for operations cost savings is a significant one, as even small efficiencies can result in significant cost savings over time. This can translate to improved profitability, increased competitiveness, and greater financial stability.

One key factor that contributes to missed opportunities in operations cost savings is the lack of effective measurement and analysis of operational data. Without a clear understanding of key performance metrics and benchmarks, it can be difficult for organizations to identify areas for improvement and implement effective cost-saving measures.

Another factor is a lack of transparency and communication between different departments and teams within an organization. This can lead to duplication of effort, unnecessary expenses, and missed opportunities to leverage shared resources or streamline processes.

In the dark fiber industry, missed opportunities for operations cost savings can come from a lack of data-driven decision-making and proper planning. The implementation of new technologies and strategies can often be delayed due to budget constraints or a lack of understanding of the potential benefits. However, the adoption of such innovations can lead to significant cost savings and improved efficiency in operations.

For example, dark fiber providers may experience inefficiencies in network management due to manual and disjointed processes. This can lead to increased downtime, poor network performance, and wasted resources. By implementing automation and artificial intelligence (AI) technologies, providers can improve network management and reduce operational costs.

Additionally, the lack of proper planning and forecasting can result in excess inventory and underutilized resources. By adopting data-driven supply chain management strategies and leveraging tools such as predictive analytics, dark fiber providers can optimize inventory levels and reduce waste, resulting in significant cost savings.

Furthermore, missed opportunities for operations cost savings in the dark fiber industry can come from a lack of communication and collaboration between different departments. For example, the installation and maintenance of fiber optic cables require coordination between the engineering, construction, and customer service departments. Without proper communication and collaboration, there may be delays, errors, and duplicated efforts. By implementing cross-functional teams and communication strategies, dark fiber providers can improve coordination and reduce operational costs.

Applying the principles of QBD can help organizations identify and capitalize on missed opportunities for operations cost savings. By better analyzing operational data, identifying key performance metrics, and implementing data-driven strategies for process improvement, organizations can significantly reduce costs and improve profitability.

In conclusion, the missed opportunity for operations cost savings is a significant one that can impact the financial health and competitiveness of organizations across all industries. By implementing effective measurement and analysis strategies, improving transparency and communication, and leveraging the power of QBD, organizations can identify and capitalize on missed opportunities for operational cost savings and improve their bottom line.

Buckle Up!

Get ready to unlock the full potential of your organization. The challenges that businesses face today are complex and ever evolving, but fortunately, you don't have to face them alone. With the QBD methodology, you have access to a powerful toolkit that can help you identify and mitigate risks while unlocking new opportunities for growth and innovation. In the next section, we'll dive deep into the four particles that make up the QBD methodology: Source, Bridge, Target, and Binary Link. We'll show you how to easily identify these particles within your existing systems, data, process, and people, and provide practical guidance on how to use them to address your unique challenges. By harnessing the power of these particles, you can take your organization to new heights and achieve unprecedented success.

SECTION I – THE FOUR PARTICLES

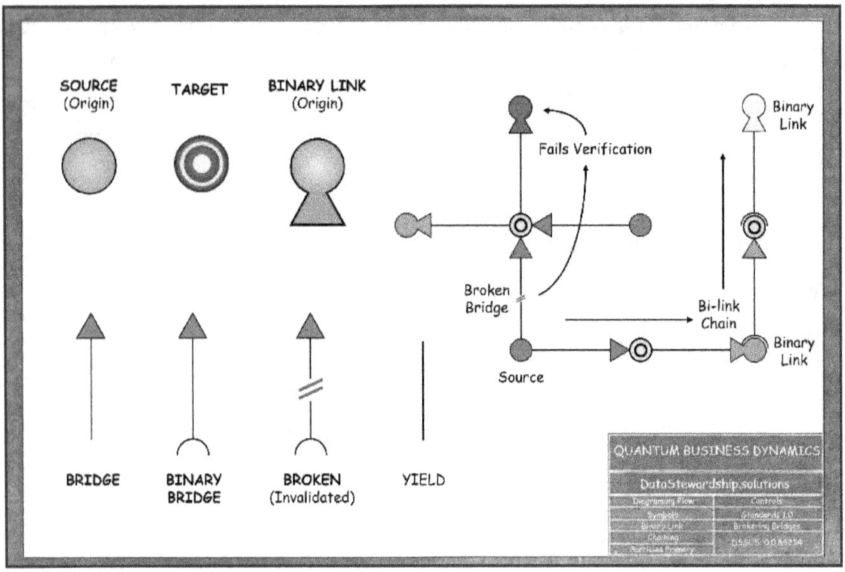

Sources

Sources are the starting point for everything in the QBD system, and they are also where we find the biggest missed opportunity for companies. According to a study by Gartner, the average company spends up to 30% of its IT budget on data integration alone (Gartner 2018). This is a clear indication of just how important data validation is to businesses. However, despite the importance of data validation, many businesses still overlook it or do not utilize it to its fullest potential.

It's ironic that businesses spend so much money and time validating their data but then fail to leverage it for further growth and efficiency. As one of our founders puts it, "Validating data is like planting a seed. If you don't cultivate it, you won't reap the rewards of a bountiful harvest."

The good news is that QBD makes it easy to capture and leverage validated data as Sources. By utilizing our easy-to-use tools, businesses can memorialize Sources within their existing systems, processes, and even employees. This can help businesses unlock new opportunities for growth and efficiency while also reducing costs and mitigating risks.

But what exactly is a Source? A Source is any person, place, or thing that is also an actionable validation that occurs as a natural part of any business's daily processes. For example, in the entertainment industry, a Source could be a copyright registration that has been validated. In the device manufacturing industry, a Source could be a validated bill of materials. In the communications industry, a Source could be a validated phone number or email address. The possibilities are endless, and every business has its own unique set of Sources.

To ensure that businesses can easily identify and utilize Sources, we have created four attributes that are easily retrievable from any software, platform, or record. These attributes are an Identifier, a Description, a Validation Type, and a relatable visual representation such as an image or a QR code. When these attributes are combined, they create the Source Particle, which can then generate additional attributes as needed.

But the benefits of Sources go beyond just validating data. They also serve as a starting point for businesses to explore new opportunities for growth and efficiency. By leveraging Sources, businesses can create new Targets for growth, such as identifying new customer segments, optimizing supply chain operations, or developing new products or services. Additionally, Sources can help businesses reduce operational costs by identifying areas where they can streamline processes and reduce waste.

It's important to note that the behavior rules for Sources are quite simple. They have no predecessor in a Binary Link and are the result of some validation step. The validation can be independently verified over and over again, and the validation can be invalidated, which will break all Bridges to Targets from that Source. Sources naturally seek Targets, and they can only point at Targets. Finally, Sources must have a user-friendly handle, which in our case is a set of 3 characters (uppercase letters or numbers), followed by a hyphen, and then followed by four more such characters (e.g., BR5-49ME) similar to old phone numbers, but each character can be alphanumeric.

One of the biggest benefits of Sources is that they are uniquely possessable, represented as digital tokens that can be withdrawn to any standard wallet that supports an underlying network. This means that they can be used across industries and even be owned by individuals, providing new opportunities for collaboration and innovation.

1. Validations

Validated processes and assets are a critical part of any organization's operations. These can include everything from validated email addresses to validated inventory items and even validated employee credentials. For businesses looking to improve their operations and increase efficiency, these validated processes and assets can provide a valuable starting point.

One example of a validated process that many businesses rely on is email validation. Every time an email is sent, it is important to ensure that the address is valid and belongs to the intended recipient. This is particularly important for businesses that rely heavily on email communication, such as marketing firms or customer service departments. By having a validated email process in place, businesses can ensure that their messages are reaching their intended audience, minimizing the risk of wasted time, resources, and potentially damaging miscommunications.

Another example of a validated process that many businesses rely on is inventory management. For businesses that sell physical products, managing inventory is a critical part of operations. Validating inventory items ensures that businesses are accurately tracking their stock levels and minimizing the risk of lost or stolen goods. This is particularly important for businesses with centralized locations, such as retail stores or warehouses, where it can be difficult to keep track of inventory in real-time.

Employee credential validation is another key aspect of many businesses' operations. Validating employee credentials is important for ensuring that workers are properly qualified for their roles. It also helps businesses comply with regulations and certifications that may be required. This is particularly important for businesses in highly regulated industries such as healthcare, finance, or education, where having properly qualified employees is critical for both legal and ethical reasons.

One example of a business that relies heavily on validated processes and assets is a pharmaceutical company. These companies must comply with strict regulations regarding drug development and testing, and the use of validated processes is critical for ensuring compliance. Validated processes are used throughout the drug development process, from initial research and development to clinical trials and final product manufacturing. By using validated processes, pharmaceutical companies can ensure that their products are safe and effective, while also minimizing the risk of costly compliance violations.

Another example of a business that relies heavily on validated processes and assets is an airline. Airlines must comply with strict safety regulations, and the use of validated processes is critical for ensuring that planes are safe to fly. Validated processes are used throughout the airline industry, from initial aircraft design and manufacturing to maintenance and repair. By using validated processes, airlines can minimize the risk of accidents and ensure that their passengers and crew are safe while in the air.

Validated processes and assets are a critical part of any business's operations. By identifying and validating these processes and assets, businesses can improve efficiency, reduce costs, and minimize the risk of compliance violations and other operational issues. Whether it is email validation, inventory management, or employee credential validation, these processes play a key role in ensuring that businesses are able to operate effectively and meet their goals.

Creating a Source particle from a validation is easy…

1. Begin by identifying any validated process or asset within your organization.
2. Create a unique identifier for each Source. This could be a simple alphanumeric code, such as a serial number or an ID number. The identifier should be unique to each Source and easy to track and manage.
3. Describe each Source in detail. Include all relevant information, such as the validation method used, the date of validation, and any other pertinent details.
4. Assign a Validation Type to each Source. This helps categorize Sources and makes it easier to identify and manage them. Validation Types could include things like Validation ID, Certifications, Memberships, a Street Address, a Confirmation, a Resource ID, and more.
5. Create a visual representation of each Source. This could be a photograph, a diagram, or any other visual aid that helps identify and manage the Source.
6. Use QBD services to memorialize each Source. This makes it easy to relate to, verify, and manage each Source.
7. Use Bridges to connect Sources to Targets. By creating Bridges, companies can unlock new opportunities for growth and success.

We can learn a great deal more about the value of Sources and how to rapidly identify them by considering the rules that govern them. Here is a list of Source Rules.

2. Sources are always a point of Origin.

The beauty of the QBD methodology is that it recognizes and leverages existing validated processes or assets within a business. These validated processes or assets are Sources that serve to start things off that typically result in value for the business. They are the origin of the chain of events and do not typically have any predecessor.

Let's consider a business that deals with shipping goods and uses a validated Bill of Lading (BOL) to ensure the shipment's authenticity and legality. The BOL validation process serves as a Source and can be used as the Origin of a Binary Link Chain. Once the BOL validation Source is established, it can be bridged with other Sources such as the Origin and destination of the shipment, the carrier, and the mode of transport. These Binary Link Chains can help businesses to track the shipment more effectively and provide real-time updates to their customers. By using the validated BOL as a Source, businesses can reduce the risk of fraud and ensure a more secure supply chain. They can also improve their customer service by providing timely and accurate information on the shipment's status.

In another scenario, a manufacturing company may have a validated process for quality control of its products. The quality control process serves as a Source that can be used as the Origin of a Binary Link Chain. This Source can be bridged with other Sources such as supplier performance data, customer feedback, and employee training records. The Binary Link Chain can then be used to improve product quality, optimize the supply chain, and enhance customer satisfaction.

Similarly, a company may have a validated process for hiring and validating employees' credentials. This process can serve as a Source that can be used as the Origin of a Binary Link Chain. It can be bridged with other Sources such as employee performance data, training records, and feedback from customers or colleagues. The Binary Link Chain can then be used to improve employee development and engagement, as well as to optimize the hiring process.

In all these scenarios, the Source serves as the Origin of the Binary Link Chain. It is the starting point from which businesses can create a chain of relationships between Sources and Targets. By leveraging existing validated processes or assets within the organization, businesses can develop more efficient and effective Binary Link Chains that can help them achieve their strategic objectives.

Moreover, the fact that Sources have no predecessor in a Binary Link Chain naturally mirrors what is already happening in typical businesses. Most businesses have existing processes or assets that are validated and serve as a starting point for their operations. The QBD methodology recognizes and leverages these validated processes or assets, allowing businesses to build on what they already have and create more value for their customers and stakeholders.

Sources are critical to the QBD methodology as they serve as the Origin of Binary Link Chains. By leveraging existing validated processes or assets within the organization, businesses can create more targeted and personalized marketing campaigns, optimize their supply chain, improve employee development and engagement, and achieve their strategic objectives. The fact that Sources have no predecessor in a Binary Link Chain naturally mirrors what is already happening in typical businesses, making the QBD methodology easy to adopt and integrate into existing operations.

3. Sources are the result of some validation step.

Validation may be performed by anyone, whether it is an outside vendor, a process or an automated system, or an employee. In any business, validation is a crucial part of the daily processes, and businesses often validate many different things as part of their operations.

For example, in the shipping and logistics industry, a Bill of Lading is a document that is used to validate the shipment of goods from one location to another. The Bill of Lading serves as a Source and can be used as the Origin of a Binary Link Chain. The validation step of the Bill of Lading ensures that the shipment is legitimate and authorized, and that the goods being transported are accurate and accounted for.

Similarly, in the healthcare industry, patient records must be validated to ensure that they are accurate and complete. This validation step serves as a Source and can be used as the Origin of a Binary Link Chain. Once the patient records have been validated, they can be bridged with other Sources such as the patient's medical history, lab results, and prescriptions. This Binary Link Chain can help healthcare providers to develop more personalized and effective treatment plans for their patients.

In the financial services industry, validation of financial transactions is critical to ensure that they are legitimate and authorized. The validation step of financial transactions serves as a Source and can be used as the Origin of a Binary Link Chain. Once the transaction has been validated, it can be bridged with other Sources such as the customer's financial history, credit score, and spending habits. This Binary Link Chain can help financial institutions to develop more targeted and customized financial products and services for their customers.

In the retail industry, validation of inventory is essential to ensure that the products being sold are accurate and accounted for. The validation step of inventory serves as a Source and can be used as the Origin of a Binary Link Chain. Once the inventory has been validated, it can be bridged with other Sources such as customer purchase history, preferences, and feedback. This Binary Link Chain can help retailers to develop more personalized and effective marketing campaigns that are likely to resonate better with their customers.

Validation is a critical part of any business, and Sources are the result of this validation step. By leveraging the QBD methodology, businesses can use these Sources as the Origin of a Binary Link Chain, which can be bridged with other Sources to develop more personalized and targeted products and services. The examples from different industries show how the validation step and the resulting Sources are already an integral part of a typical business, and how they can be leveraged to achieve greater success and innovation.

4. The validation and its relationship to Targets can be independently verified over and over again.

The relationship between a Source and a Target can be independently verified multiple times to inexpensively ensure its accuracy and authenticity. This verification process is essential because it guarantees that the information derived from the Source is reliable without the need for repeated more costly validations, and any decisions or actions made based on that information are accurate.

For example, consider a business that relies on inventory management. The business validates inventory items to ensure that they are accurately recorded in the system, and any discrepancies are promptly identified and addressed. But what if every time a new question regarding that inventory came up the business paid for a full new validation, their costs would skyrocket.

Similarly, in the case of a digital marketing campaign, the verification of the Source and its relationship to the Target is crucial. If the Source data used to create the campaign is inaccurate or outdated, the campaign may not achieve the desired results. This can lead to wasted resources and lost opportunities. But once validated, it would be a waste if one week later they decided to run another campaign but stopped to pay for a complete revalidation first.

Re-use was a huge buzz word just a few years ago, and yet we are dropping validations on the floor and not leveraging their re-use. Ask yourself this question, "what other groups in my organization, and what other businesses we regularly work with might pay to verify this validation to avoid paying for it themselves?" If you were to charge (internally or externally) a 20% fee for each verification, saving your business customer 80% by not having to validated the Source themselves, then all you would need is to verify 6 times to actually make profit from your validation costs.

That's a game-changer.

5. The validation can be invalidated, which will break all Bridges to Targets from that Source.

The ability to invalidate a validation may seem counterintuitive at first glance, but it is a crucial aspect of the QBD methodology. Invalidating a validation means that the validation step that created the Source is no longer valid or reliable. When this happens, all Bridges from that Source to any Targets are immediately broken and all downstream values are notified that they are now suspect.

This feature is important for a few reasons. First, it ensures the accuracy and reliability of the Sources and the resulting Binary Link Chains. If a validation step is no longer valid, it could lead to inaccurate data and ultimately harm the business. By breaking all Bridges from that source, the business can avoid any potential harm from inaccurate data.

Second, the ability to invalidate a validation allows for flexibility and adaptability in the business. As the business landscape changes and new information becomes available, it may be necessary to revalidate certain Sources. By invalidating the previous validation, the business can easily update its Sources and Binary Link Chains to reflect the new information.

For example, consider a business that uses customer feedback to create Sources and Binary Link Chains. If the business receives negative feedback from a customer and later finds out that the feedback was given in error, they can invalidate the previous validation and break any Bridges to Targets based on that feedback. This allows the business to maintain the accuracy and reliability of their data and make informed decisions based on accurate information.

Overall, the ability to invalidate a validation is an important feature of the QBD methodology that allows businesses to ensure the accuracy and reliability of their data while also maintaining flexibility and adaptability in a constantly changing business landscape.

6. Sources naturally seek Targets.

The nature of a Source is such that it naturally seeks out a Target. This is because the validation process that creates a Source is usually done with a specific goal in mind, and that goal often involves a corresponding Target. For example, in the architecture and engineering industry, a common example of Sources naturally seeking Targets is when an engineer validates an architect's blueprints. The engineer's validation serves as a Source, while the Target is the physical construction of the building. Once the blueprint validation Source is established, it can be bridged with other Sources, such as validating the materials used in construction or the dimensions of the building components. These Binary Link Chains can help architects and engineers to build better structures by ensuring that each component of the building is validated and verified before construction begins. The blueprints themselves, which serve as the initial Source, naturally seek out the Target of a better building.

The natural inclination of a Source to seek out a Target is what makes the QBD methodology so effective. By recognizing and utilizing this innate behavior, businesses can more easily identify and create new connections between Sources and Targets, leading to increased efficiency and productivity.

Moreover, because Sources can only point to Targets through Bridges, this natural inclination also ensures that those Bridges are always purposeful and intentional. By focusing on building Bridges that connect Sources to specific Targets, businesses can avoid wasting time and resources on connections that don't serve a clear purpose.

In essence, the natural seeking behavior of Sources allows businesses to be more strategic and intentional in their efforts to create value through the QBD methodology. By identifying and validating the right Sources, and then creating purposeful Bridges to their corresponding Targets, businesses can unlock new opportunities for growth and success.

7. Sources can be verified over and over again.

In today's fast-paced business environment, companies are constantly validating various aspects of their operations. From verifying the authenticity of customer information to confirming the accuracy of financial records, validation is a necessary and often costly part of doing business. However, despite the amount of effort and resources that go into these validation processes, many businesses fail to take advantage of the data and insights that can be gleaned from them.

For example, consider an engineering firm that regularly validates the blueprints and designs provided by architects for new construction projects. This validation process is crucial to ensure that the final product meets safety regulations and other requirements. However, once the validation is complete, many companies simply move on to the next step without fully utilizing the data they have gathered.

Instead, by capturing and storing this validation data, companies can create a more efficient and cost-effective verification process. For instance, the engineering firm in the example above could store information about the validated blueprints in a centralized database that could be accessed by all relevant parties. This would not only reduce the need for redundant validation steps but would also provide a valuable resource for future projects.

Despite the benefits of this approach, many businesses continue to operate in a validation-rich and verification-poor environment. This can lead to unnecessary costs, lost productivity, and missed opportunities for growth and innovation. By recognizing the value of the validation data, they are generating and taking steps to capture and utilize it, companies can gain a competitive advantage and position themselves for long-term success.

In short, businesses must shift their focus from simply validating processes to leveraging the data and insights generated by these validations.

By doing so, they can streamline their operations, reduce costs, and create a more efficient and effective verification process. In today's competitive business environment, the ability to capture and utilize validation data is essential for success.

8. Sources can only point at Targets.

In other words, once a Source has been validated and established, it can be linked with a target, which is the desired outcome of the validation process.

It's important to note that a Source can only point to a Target and not vice versa. This means that the Target cannot be used as a starting point of a Binary Link Chain, and it can only be linked with an Origin, which can be a Source that has already been established and validated. This is because the validation process is necessary to ensure the accuracy and reliability of the information that is being used to establish the Binary Link Chain.

For example, consider a dental clinic that wants to improve its tooth repair procedures by validating its patient's dental x-rays. The dental x-ray validation process serves as a Source and can be used as the Origin of a Binary Link Chain. Once the dental x-ray Source is established, it can be bridged with other Sources such as a patient's medical history, allergies, and previous dental work. These Binary Link Chains can help the dental clinic to develop a more targeted and personalized treatment plan that is likely to provide better results for the patient.

In this example, the dental x-ray is the source, and it can only point to its corresponding Targets such as the patient's medical history, allergies, and previous dental work. The Targets cannot be used as a starting point of the Binary Link Chain because they do not have any validation process associated with them. Therefore, it's important for dental clinics to establish and validate their Sources, such as the dental x-rays, before linking them with Targets to ensure the accuracy and reliability of the information that is being used to drive their tooth repair procedures. By doing so, the dental clinic can ensure that they have accurate and up-to-date information about their patient's dental health, which can help them make better decisions about their treatment plans.

9. The only way for a Source to be related to a Target is by a Bridge.

In the QBD methodology, the only way for a Source to be connected to a Target is through a bridge. This means that in order for a Source to provide value to a business, it must first be linked with a Target that is relevant to the business's goals and objectives.

The Bridge between the Source and Target can be created in many ways. It could be a manual process such as an employee entering customer feedback information into a connected CRM system and linking it with the customer's purchase history. Or it could be an automated process where the customer feedback is automatically linked with the customer's purchase history based on certain criteria such as the customer's name or email address.

Regardless of the form the Bridge takes, it is crucial for businesses to establish clear and reliable Bridges between their Sources and Targets to maximize the value that can be derived from the QBD methodology. By doing so, businesses can gain a deeper understanding of their operations, customers, and market, and make more informed decisions that can drive growth and profitability.

10. Sources must have a unique user-friendly handle.

Sources must have a user-friendly id or handle, which serves as an identifier to identify and reference them quickly and easily. In the case of QBD, this handle is a unique set of 3 characters (uppercase letters or numbers), followed by a hyphen, and then followed by four more such characters (e.g., BR5-49ME) that is similar to old phone numbers, but each character can be alphanumeric. This allows for a large number of unique handles to be generated, making it easy to create and manage multiple Sources without the risk of duplication or confusion.

For example, a healthcare provider may use this user-friendly handle to track patient records. Each patient would have a unique handle assigned to them, allowing for quick and easy identification of their medical history and any ongoing treatments or medications. This handle can also be used to link with other Sources such as insurance information or appointment schedules, creating a comprehensive record of the patient's healthcare information.

Similarly, a retail business may use this handle to track inventory items. Each item would have a unique handle assigned to it, allowing for quick and easy identification and tracking of stock levels, pricing, and supplier information. This handle can also be used to link with other Sources such as sales history or customer feedback, providing insights into which products are popular with customers and which may need to be restocked or discontinued.

Overall, the use of a user-friendly handle for Sources is an important aspect of QBD as it allows for efficient management and referencing of Sources within a business's ecosystem.

11. When created, Sources naturally generate new attributes.

Once a Source is created and validated, it generates new attributes that can be used to better understand and manage the business processes. These attributes include confidential and public information.

The confidential attributes of a Source include the user-friendly handle, the Identifier, the Description, and the present state (valid/invalid), and can also have additional attributes. These attributes are kept private and are only shared when there is a need and authorization to do so. For example, an employee's validated credentials may be kept confidential until the need arises to share them with a client or regulatory agency.

On the other hand, the public attributes of a Source, while discernible from the public record, contain minimal information. For example, a validated inventory item may have a public image that can help identify it, as well as a CID (content identifier) hash that points to an immutable public business object, a Series identifier, and a Serial Number. But these public attributes cannot be used to draw relationships to other particles, nor track the item's movement throughout the supply chain.

In this way, the private things stay private, while the attributes that are necessarily public are public.

It's important to note that no other data needs to be disclosed beyond the public attributes of a Source. This ensures that sensitive information remains confidential and is only shared when necessary. By utilizing these attributes, businesses can better manage their processes, improve transparency, while protecting sensitive information.

When a Source is validated, it generates new attributes that can be used to better understand and manage business processes. These attributes include confidential and public information that is kept private or discernible from the public record. By utilizing these attributes, businesses can improve transparency and protect sensitive information while also improving their efficiency and accuracy in business processes.

12. There's no way to suss out a Source's relationships or even that it is a Source from publicly available data because all particles share the exact same public attribute footprint.

This aspect of the QBD methodology is important for maintaining privacy and security within a business's network of Sources and Targets. By using a standardized public attribute footprint for all particles, there is no way for outside parties to determine the relationships between Sources and Targets or to gain information about the nature of the business's operations.

For example, consider a business that handles sensitive financial information for its clients. By keeping the relationships between Sources and Targets private, the business can ensure that its clients' information remains secure and confidential. Even if a hacker were to gain access to the business's system, they would not be able to determine which Sources correspond to which Targets, making it much more difficult to compromise the security of the system.

At the same time, the standardized public attribute footprint also makes it easier for businesses to share information with each other while maintaining privacy and security. Because all particles share the same public attribute footprint, businesses can easily establish connections and share data without compromising the privacy of their own internal networks.

Overall, the ability to maintain privacy and security while still sharing information is a crucial component of the QBD methodology. By standardizing the public attribute footprint for all particles, businesses can protect their own sensitive information while still collaborating and sharing data with other businesses.

13. The record of their attributes is immutable and permanent, good for reproof, verification, and even legal nonrepudiation.

The immutability and permanence of the record of Source attributes is a key feature of the QBD methodology. This means that once a Source is validated and its attributes recorded, they cannot be changed or tampered with. The record of the source's attributes serves as proof of its validation and can be used for reliable verification purposes. This is particularly useful in situations where there is a dispute over the validity of a Source or its relationship to a Target.

Furthermore, the immutable record of Source attributes can also serve as a form of legal nonrepudiation. Nonrepudiation is a legal concept that refers to the inability of a party to deny the authenticity or integrity of a communication or transaction. In the context of QBD, the immutable record of a source's attributes can be used as evidence in legal disputes to prove the validity of a transaction or communication.

For example, consider a business that uses the QBD methodology to validate its contracts with suppliers. Each contract is recorded as a Source with its attributes, including a unique identifier, the date of validation, and the parties involved. If a dispute arises between the business and a supplier over the terms of the contract, the immutable record of the contract's attributes can be used as evidence in legal proceedings to prove the validity of the contract and the parties' agreement to its terms.

In addition to its use in legal disputes, the immutable record of source attributes can also be used for internal auditing purposes. By reviewing the records of validated Sources, businesses can ensure that their processes and systems are functioning correctly and that their data is accurate and up to date. This can help to identify areas for improvement and increase overall efficiency and effectiveness.

Overall, the immutability and permanence of the record of source attributes is a valuable feature of the QBD methodology. It provides a reliable and verifiable record of validation that can be used for verification, legal nonrepudiation, and internal auditing purposes.

14. Verifications of any of the attributes also become immutable and permanent, good for reproof, verification, and even legal nonrepudiation.

Verifying the attributes of a Source is an important aspect of the QBD methodology. It allows businesses to ensure the accuracy and reliability of the information that they are using to make decisions. Verifying the attributes of a Source involves checking the information that is associated with it. For example, if a business is using a validated email address as a Source, they may need to verify the owner of the email address, the date of validation, and any other relevant information that is associated with it.

One of the key benefits of verifying the attributes of a Source is that it creates an immutable and permanent record of the verification, itself. This subsequent verification record can then be used itself for reproof, verification, and even legal nonrepudiation. If there is ever a dispute about the verification, say from regulatory agents, the record of the verification can be used to settle the matter.

For example, let's say that a business uses a validated certification required by regulations. If there is ever a dispute over compliance, the record of the verification can be used to prove that the valid certification was indeed verified by the business. This can help to settle any disputes and ensure that the business is compliant with regulations.

15. Sources are singular unto themselves and there cannot be a copy of them.

Sources are unique entities within the QBD framework and cannot be replicated nor duplicated. Each Source represents a specific validation process or asset, and its attributes are unique to that specific validation. This ensures that all of the Bridges stemming from a particular actionable validation each point back to only that unique Source. Thus, each related Binary Link Chain is based on accurate and reliable information.

For example, let's say a school has a validated student credential as a Source. This Source cannot be duplicated or replicated, as each student has a unique credential that represents their specific qualifications and skills. If a duplicate Source were created, it would not accurately represent the student's credential, and could potentially lead to incorrect or unreliable information being used in business decisions.

By ensuring that each Source is singular and unique, businesses can have confidence in the accuracy and reliability of the information used in their Binary Link Chains. This also helps to prevent data duplication and inconsistencies, as each Source is only used once within the QBD framework.

16. Sources are uniquely possessable, as they are represented as digital tokens that can be withdrawn to any standard wallet that supports an underlying network.

Sources in QBD are uniquely represented as digital tokens, which means they can be transferred and owned in a secure and transparent manner. The tokenization process involves converting the source's attributes into a unique digital representation that can be stored on a blockchain, hashgraph, or other secure network. This allows for easy transfer and management of ownership and usage.

Since Sources are uniquely possessable, they can be withdrawn to any standard wallet that supports the underlying network. This means that businesses can easily transfer ownership of a Source to another entity, such as a partner or vendor, without the need for complex legal agreements or paperwork. It also allows for easy tracking of ownership and usage, ensuring that Sources are used appropriately and for their intended purpose.

Tokenization also provides an added layer of security, as the digital tokens are cryptographically secured and can only be accessed by authorized parties. This helps to prevent fraud, tampering, and other malicious activities that can compromise the integrity of the source's attributes.

Furthermore, the use of digital tokens allows for seamless integration with other blockchain-based technologies, such as smart contracts. Smart contracts are immutable self-executing contracts with the terms of the agreement directly written into code, which can automatically trigger actions based on predefined conditions. By integrating Sources with smart contracts, businesses can automate processes and ensure that the source's attributes are only used in accordance with the agreed-upon terms.

17. Sources may only have one Validation Type assigned to them.

Sources can have only one type assigned to them, such as an email address, a phone number, a street address, or a resource ID. The type assigned to a Source reflects the nature of the validation process that created it. For example, a Source with the Validation Type "Validation ID" might be created through a process that verifies a person's identity, while a Source with the Validation Type "Certification" might be created through a process that validates a person's skills or credentials.

As businesses identify new types of validation processes, they can create new Validation Types for Sources that reflect those processes. This enables them to create a more comprehensive and detailed network of Sources and Targets, which in turn can lead to more accurate and effective decision-making.

The number of Validation Types available for Sources is expected to grow rapidly as businesses continue to innovate and improve their validation processes. Some examples of Validation Types that may be added in the future include memberships, confirmations, and resource IDs. Anything that can be validated can be assigned to an existing type or a new type can be created for it.

By assigning a Validation Type to a source, businesses can easily categorize and manage their Sources, which can help them to identify and resolve issues more quickly. It also makes it easier to identify the relationships between Sources and Targets, which is crucial for effective decision-making.

18. A Sources Identifier must be unique.

The unique Identifier of a Source is a critical component of the QBD system. It ensures that each Source can be easily identified and distinguished from others. The Identifier is typically a combination of alphanumeric characters that serve as a unique digital signature for the Source. It must be unique to prevent any confusion or ambiguity when searching for or linking to a specific Source.

For example, let's say a law enforcement agency uses the QBD system to validate evidence as it is being curated at a crime scene. Each item's record can be established as a separate Source with a unique Identifier. This allows the law enforcement agency to easily distinguish between the different pieces of evidence and link them with corresponding Targets, such as investigations, court cases, or appeals.

For the system to work effectively, it's important that each Source has a unique Identifier. If two Sources share the same Identifier, it can cause confusion and potentially result in the wrong Source being linked with a target. Additionally, duplicate Identifiers can also make it difficult to search for specific Sources, as multiple Sources may appear in search results when only one is correct.

Therefore, it's important for businesses to establish a clear system for assigning unique Identifiers to their Sources. This may involve creating a standard naming convention or using an automated system to generate unique Identifiers. Regardless of the approach, it's critical to ensure that each Source has a unique Identifier to maintain the accuracy and reliability of the QBD system.

Sources in Summary

After exploring the concept of Source particles in-depth, it becomes clear that they are an integral component of the Quantum Business Dynamics™ (QBD™) methodology. Sources serve as the starting point or Origin of Binary Link Chains and are created through a validation process. They can only point to Targets and their relationships can only be established through Bridges. Sources are also uniquely possessable and have a user-friendly handle that is immutable and permanent, making them ideal for verification and nonrepudiation purposes.

Businesses validate a multitude of things on a daily basis, and these validations serve as the foundation for creating Source particles. By leveraging these validated Sources, businesses can develop targeted and personalized marketing campaigns, make better decisions about purchasing, production, and sales, and improve their overall supply chain management.

It's also important to note that while Sources are rich in validation, businesses are often verification poor. They validate the same things repeatedly without subsequently using the validated information for verification purposes. This is a costly waste that can be addressed through the use of Source particles and the QBD methodology.

Sources must have a unique identifier and can only be related to Targets through Bridges. They are represented as digital tokens that are uniquely possessable and have one assigned type. As businesses continue to identify new things that can be validated, the number of types is expected to grow rapidly.

Overall, Source particles are a critical component of QBD that can help businesses increase efficiency and improve decision-making. By establishing validated Sources and linking them with Targets through Bridges, businesses can develop a more complete understanding of their operations and better position themselves for success.

In the next section, we will explore the concept of Targets and their role in the QBD methodology. We will delve into what makes a good target, how to set them out to attract Sources, and how to bridge your business to other businesses using QBD fabrics.

Targets

Targets are the end goal that any business or organization is working towards. It is a relatable desire that is always present in any company's daily processes. According to Forbes, "Without goals, organizations don't know what they're working towards, and they don't know when they've arrived." Businesses set targets to drive growth, create new opportunities, and ultimately increase their bottom line.

Targets are as diverse as the industries they serve. For example, a pharmaceutical company may have a target to develop a new drug that meets a specific medical need, while a retail company may have a target to increase sales by a certain percentage. Whatever the target may be, it is crucial to define it well to create value and avoid risks and liabilities. Failing to properly identify or define targets can lead to a lack of direction and wasted resources.

One of the most exciting aspects of Targets is that they naturally draw Sources towards them. When a company identifies a target, they can then begin to collect and bridge Sources that will help them achieve that target. As businesses continue to create Targets and bridge Sources, the resulting Bi-link Fabric becomes a valuable asset that can be used for auditing, compliance, and subsequent verification.

Targets are created by someone with the role of Authority, who combines six key attributes that are easily retrievable from any software, platforms, records, or even employees. These attributes include an Identifier, a

Description, Agency, Allowable Validation Types, What Role Pays for Verification, who will end up as Owner for the subsequently yielded Binary Links created from agreements to this Target, and a voluntary list of data partners. Targets generate new attributes beyond the original six upon their creation, including public and confidential attributes. These attributes provide a comprehensive record of the Target and its relationships to other particles within the Bi-link Fabric.

The Target Identifier must be unique, and like Sources, they are singular unto themselves, and there cannot be duplicates. Targets do not have types, but they can specify the types of Sources that may be bridged to them from a multi-select of all Validation Types. Targets may either have unlimited Bridges or may set a specific Bridge limit. When created, Targets can generate new attributes beyond the original six that caused their creation.

Targets cannot be linked directly to a source; they must be linked through a Bridge that has been authorized to do so. This Bridge creation between an Origin and a Target then generates a Binary Link downstream of the target. Targets do not have agency, but they do set the agency for all Binary Links created through them. Targets must remain in the treasury of the Governor of the platform and are not possessable, ensuring the integrity and validity of the Bi-link Fabric created by all businesses.

Targets are the driving force behind any business or organization. They are the end goal that drives growth, creates new opportunities, and ultimately increases the bottom line. Creating and defining Targets well can lead to significant value and opportunities for an organization, failure to identify them leads to risks and liabilities. The unique attributes of Targets, along with their relationship to Sources and the resulting Bi-link Fabric, make them an essential part of any business.

1. Targets can have a predecessor, which is Bridge from an Origin. Origins can be either Sources or Binary Links.

Targets are the desired outcomes that businesses strive to achieve through their operations. These outcomes can range from sales targets to quality targets to customer satisfaction targets. To achieve these targets, businesses must identify the necessary resources and inputs, which can come from Sources or Binary Links. Bridges are used to connect these inputs to the desired outcomes, creating a clear path for achieving the target.

Targets do not necessarily need to have a predecessor in the form of a Bridge from an origin. While it is true that Targets can be linked to Sources or Binary Links through Bridges, Targets can also be stand-alone goals or objectives that a business aims to achieve.

For example, a business may have a target to increase sales by 20% in the next quarter. This target does not necessarily have a direct Bridge from an origin, but rather is a standalone goal that the business aims to achieve through various actions and strategies.

That being said, Targets can often be more effectively achieved by linking them to relevant Sources or Binary Links through Bridges. This helps ensure that the business is using accurate and validated information to inform their decisions and actions towards achieving the target.

The importance of this feature is that it allows businesses to clearly define and measure their progress towards achieving their goals. As noted by entrepreneur and author Steve Blank, "A startup is a temporary organization designed to search for a repeatable and scalable business model." By setting Targets and identifying the necessary inputs, businesses can more effectively search for and refine their business models.

Furthermore, connecting Targets to Origins through Bridges helps to ensure that the necessary resources are being used in the most efficient and effective way possible. As noted by management consultant Peter Drucker, "Efficiency is doing things right; effectiveness is doing the right things." By bridging the inputs to the desired outcomes, businesses can ensure that they are not only doing things right but also doing the right things.

Research has shown that businesses that set clear and measurable targets are more likely to achieve their goals. According to a study by Harvard Business Review, "Leaders who are most effective at driving growth and creating long-term value set clear and measurable targets, engage the organization to achieve them, and hold themselves accountable for results." By clearly defining their targets and bridging the necessary inputs, businesses can increase their likelihood of success.

In summary, the ability for Targets to have a predecessor through Bridges from Origins is a critical feature that enables businesses to identify and achieve their desired outcomes. By connecting the necessary inputs to the desired outcomes, businesses can ensure that they are doing the right things in the most efficient and effective way possible.

2. Targets are created by someone with the role of Authority and are the result of business analysis.

Targets are not created arbitrarily or without purpose. They are the product of careful analysis and planning, with the ultimate goal of achieving some desired outcome or improving some aspect of the business. The creation of Targets is typically the responsibility of individuals in Authority roles within an organization, who have the necessary expertise and knowledge to identify and prioritize the most important goals for the business.

According to a report by McKinsey, effective target setting can improve business performance by up to 25%. This underscores the importance of careful analysis and planning when creating Targets. The report further states that targets should be specific, measurable, attainable, relevant, and time-bound (SMART), as this increases the likelihood of achieving the desired outcomes.

The creation of Targets is an ongoing process, as businesses continually reassess their goals and objectives in response to changing market conditions, customer needs, and other factors. For example, a business may set a Target to increase sales by 10% in the next quarter, but if market conditions change, the Target may need to be revised or updated.

Targets are created by someone with the role of "Authority" and there should be many people within your organization with that role. According to a study by Harvard Business Review, involving employees in the target-setting process can lead to better performance and increased buy-in from employees, as they feel more invested in the success of the organization.

Overall, the creation of Targets is a critical aspect of business planning and strategy and should be approached with care and attention to detail. By setting specific and measurable Targets, businesses can improve their performance and achieve their desired outcomes.

3. Targets can be part of a verification from an Origin, but that verification will never reveal other Origins, neither downstream nor upstream via other Bridges from other Origins.

Targets are designed to be specific and focused on achieving a particular business goal. The creation of a Target is a deliberate process that involves careful analysis and planning. The authority responsible for creating the Target needs to have a clear understanding of the business's needs and objectives. This can involve analyzing the company's products, systems, and services to identify areas where improvements can be made.

Once a Target has been established, it serves as a guidepost for the business. It provides a clear direction for the company and helps to ensure that everyone is working towards the same goal. In some cases, Targets can also be used as part of a verification process. This involves checking that a particular Source has been validated and is suitable for use in achieving the target.

However, it's important to note that a verification process will never reveal other Origins or Bridges from other Origins. The scope of any one verification for a specified Source will terminate at the resulting Binary Link yielded on the other side of the target. This is an important aspect of the Target particle as it ensures that information is not shared unnecessarily and that each Source and Binary Link remains separate and distinct.

In the context of data privacy and security, this feature is particularly important. By limiting the scope of verifications to a particular target, businesses can ensure that sensitive information is not inadvertently shared. It also helps to prevent data breaches and unauthorized access to information.

Overall, the creation of a Target is an important part of any business strategy. It provides a clear direction and helps to ensure that everyone is working towards the same goal. Additionally, the limitations on verifications ensure that sensitive information is protected and that each Source and Binary Link remains distinct.

4. Targets can have limits to the number of Bridges allowed to them or allow an unlimited number of Bridges.

The ability to limit the number of Bridges allowed to a Target is a useful feature that allows businesses to control access to resources and prevent overuse. This can be seen in various industries, such as entertainment licensing where a certain number of viewers are permitted. By setting a limit on the number of Bridges allowed, businesses can avoid unauthorized access and ensure that the resource is being used appropriately.

On the other hand, there are Targets that allow for an unlimited number of Bridges. This is often the case in marketing and advertising, where businesses want to reach as many potential customers as possible. For example, a company might create a Target to attract potential customers who are interested in a certain type of product. They may use various Bridges, such as social media, email marketing, and online advertising, to reach as many people as possible.

Generally speaking, businesses with unlimited Targets and Bridges are more likely to be successful in achieving their marketing goals. Our experience shows that businesses that used multiple Bridges to reach their Target audience had a higher conversion rate than those that used only one Bridge. This highlights the importance of having the flexibility to use multiple Bridges to achieve a Target's goals.

However, it is important to note that having too many Bridges can also be overwhelming and difficult to manage. Businesses need to strike a balance between the number of Bridges used and their effectiveness in achieving the Target. As such, businesses should carefully consider their Targets and the Bridges that they use to achieve them to ensure that they are maximizing their resources and achieving their goals.

5. Targets naturally draw Sources.

Targets are created with the purpose of drawing in specific Sources, which can be validated data, information or even entities. According to research, businesses that are able to effectively align their Targets with the appropriate Sources tend to see better results in achieving their goals. For example, if a company sets a target to increase sales of a particular product, they can use Sources such as customer feedback, sales data, and market research to better understand their target audience and develop a marketing strategy that resonates with them. In this way, the Sources are naturally drawn to the Target, as they are all working towards achieving the same goal.

Furthermore, Targets can be used to drive innovation within a business. By setting ambitious targets, companies can push themselves to explore new ideas and approaches, leading to the development of new products, services, and processes. According to a study by McKinsey & Company, companies that set ambitious targets tend to see higher revenue growth than those that do not. The study found that "the top-performing quartile of companies set targets that were 60 percent higher than the rest."

In order to effectively draw in Sources, it's important for businesses to have a clear understanding of their targets and the types of Sources that are most likely to help them achieve their goals. This requires a deep understanding of the market, the target audience, and the competitive landscape. By leveraging data analytics and other tools, businesses can gain insights into these factors and develop strategies that are more likely to succeed.

6. Targets set their own sense of agency.

The concept of agency is crucial in understanding the relationship between Targets and Sources. In different industries, the role of agency may vary, but it remains the key mediator in managing the relationship between the two. As mentioned in the previous point, Targets can have limits to the number of Bridges allowed to them, and this is where the role of agency comes in. The agent acts as a manager of a Bridge to manage the relationship between the Source and the Target, ensuring that the limits set by the Target are not exceeded.

In some industries, like real estate or entertainment, the role of agency is well-defined, and the person filling the role is often referred to as an agent. However, in other industries, like the military, the role of agency may not be as clear-cut. For instance, when a validated weapon is a Source, and a specific military action is the Target, the agency may be passed from the Navy to the Marines, and ultimately the Army, as the weapon moves from a ship at sea to the actual engagement on land.

Furthermore, agency is portable for a single fixed, immutable Source bridged to a Target. An example of this is in the telecommunications industry with Local Number Portability. This allows customers to keep their phone number and access to the GSM network when they switch carriers without notifying their carrier or asking for permission. The mechanism that ensures this portability is known as agency portability.

It's important to note that the Target particle dictates the agency for all Binary Links created through it. The agent has a critical role in managing the relationship between the Source and the Target, ensuring that the limits set by the Target are adhered to, and that the Binary Links are created within the bounds of the agency. In summary, agency is specific to an industry and is an essential aspect of managing the relationship between Targets and Sources, ensuring that business goals are met, and compliance with limits and regulations is maintained.

7. When an Origin is bridged to a Target, the relationship yields a downstream Binary Link.

Targets serve as the end goal of a Bridge and can only be linked downstream to Binary Link tokens, which are automatically generated when an Origin (Source or another Binary Link token) is bridged to it. The resulting Binary Link token represents the relationship between the Source and the Target, and the Target can only relate to this Binary Link token through its yield.

The yield does not have an identifier. It is the process by which a Binary Link token is created from the bridging of an Origin to a Target. This Binary Link token is a possessable particle that contains information about the bridged relationship, including the validation of the Origin and any other relevant attributes. The Binary Link also serves as a verification record that can be used for auditing or compliance purposes, as well as subsequent verification.

This downstream linking ensures that the integrity of the relationship between the Source and the Target is maintained and that the resulting Binary Link token is securely stored and accessible only to authorized parties. This allows businesses to better manage their data and ensure that it is being used appropriately, while also reducing the risk of data breaches or other security issues.

The use of Binary Link tokens also allows for easier tracking and analysis of data, as each relationship between a Source and Target can be easily identified and analyzed. This can help businesses to better understand their processes and make more informed decisions based on the data that is available to them.

Overall, the downstream linking of Targets to Binary Link tokens helps to ensure the accuracy and reliability of the information that is being used to drive business decisions, while also improving data security and facilitating better data analysis.

8. Targets can only be linked upstream to Origins that have been bridged to them.

Targets are an integral part of a business's processes, as they represent specific goals or outcomes that are desired. In order to achieve these goals, businesses must identify the appropriate Sources that will help them reach their Targets. However, there is a critical step that must occur before any linkage can be made between a Source and a Target: the creation of a Bridge. The Bridge serves as the relationship between the Origin and the Target, and any upstream linkage from the Target must be made through one.

9. Targets can be transferred from the Authority to another party, who then takes on the role of Authority going forward.

The dynamic nature of businesses and their relationships necessitates flexibility in the management and control of Targets within the QBD system. While the Authority initiates and controls a Target, there are scenarios where this Authority may need to be transferred to another party. This transfer effectively passes the mantle of control and responsibility for the Target from one party to another.

Imagine a situation where a Target is a project within a company, and the Authority is the project manager. Should the project manager leave the company or be reassigned to a different project, the Authority over that Target could be transferred to a new project manager. Similarly, in more complex scenarios such as mergers or acquisitions, the Authority over certain Targets might need to be transferred from one organization to another.

It's crucial to note that this isn't a simple change in labels. The party receiving the transferred Target becomes the new Authority and assumes all the associated responsibilities. They control the interactions with the Target, including what Origins can bridge to it, and they oversee the continuing evolution of its associated Bi-link Fabric.

As with many aspects of QBD, this feature underscores the flexibility and adaptability of the system to the ever-changing dynamics of business environments. It's a testament to QBD's potential for maintaining a robust, yet flexible, structure for data management, even amidst changes in roles and responsibilities.

Targets can only be linked upstream to Origins that have been related to them by a Bridge. This means that a Target cannot be linked to a Source directly, but rather it must be linked through a Bridge that has been requested by some system or person that has authorization to do so. The Bridge creation between an Origin and a Target then generates a Binary Link downstream of the Target. This is a crucial aspect of the relationship between Targets and Bridges, as it ensures that any verification or auditing can be traced back to the original Origin.

This linkage through a Bridge serves to maintain the integrity and validity of the relationship between the Source and the Target. By requiring a Bridge, any upstream linkage from the Target can be traced back to the original Origin, and any verification or auditing can be validated through the Binary Link generated downstream of the Target. This ensures that any data or information used to drive business decisions is accurate and reliable.

For a Bridge to be created, some system or person with authorization must request it. This request is typically made by someone in a position of Authority within the business, such as a manager or administrator. Once the Bridge has been established, any upstream linkage from the Target must be made through it. This helps to maintain the integrity and validity of the Bridge and ensures that any verification or auditing can be traced back to the original Origin.

The requirement of a Bridge also helps to prevent unauthorized linkage between a Source and a Target. By requiring a Bridge, any linkage must be authorized by someone with the appropriate permissions, which helps to prevent any unauthorized access to sensitive data or information. This is particularly important in industries where data privacy and security are critical, such as healthcare and finance.

Finally, Targets can only be linked upstream to Origins that have been related to them by a Bridge. This requirement helps to maintain the integrity and validity of the relationship between the Source and the Target and ensures that any data or information used to drive business decisions is accurate and reliable. By requiring a Bridge, any upstream linkage from the Target can be traced back to the original Origin, which helps to prevent unauthorized linkage and ensures that any verification or auditing can be validated through the Binary Link generated downstream of the Target.

10. Targets, like Sources, must have a unique user-friendly handle.

Just like Sources, Targets must also have a user-friendly handle that can be easily remembered and recognized. The handle is a unique identifier that distinguishes the Target from other Targets and helps to establish a clear association between the Target and its related Binary Links. The handle is composed of a 3 character-4 character pairing, similar to old phone numbers, where each character can be alphanumeric.

The user-friendly handle is an essential attribute of the Target particle, as it helps to facilitate communication and recognition between different parties that are involved in the validation process. It provides an easy way to identify and refer to a specific Target and helps to prevent confusion or errors that can arise from using complex or ambiguous identifiers.

One example of how the user-friendly handle can be used in a business context is in supply chain management. A company may create a Target for a specific inventory item that they need to track throughout the supply chain. The user-friendly handle can be used to easily identify and refer to the inventory item and its related Binary Links, such as supplier information, delivery schedules, and sales history.

Moreover, the user-friendly handle is also useful for security purposes, as it allows for quick identification and tracking of any suspicious activity or breaches that may occur in the validation process. By having a unique and easily recognizable handle, businesses can more effectively monitor and respond to any potential security threats that may arise.

In addition, the use of alphanumeric characters in the user-friendly handle allows for greater flexibility and creativity in choosing handles that are both memorable and unique. This can help businesses to create a more distinct brand identity and establish a stronger connection with their customers.

Overall, the user-friendly handle is an important attribute of the Target particle that plays a crucial role in facilitating communication, validation, and security in a variety of business contexts. By providing a unique and easily recognizable identifier, it helps to establish clear associations between Targets and their related Binary Links and provides an essential foundation for the validation process.

11. When created, Targets naturally generate new attributes beyond the original six that caused their creation.

When a Target is created, it generates new attributes beyond the original six that were used to create it. These attributes are important for maintaining the integrity and validity of the Target as well as providing additional information about the Target.

The confidential attributes of a Target include the user-friendly handle, the Identifier, the Description, their present state (valid/invalid), their agency, allowable Validation Types, what role pays for related verifications involving the Target, and a list of data partners voluntarily set by the Authority. These attributes are kept private and can only be accessed by authorized personnel or systems. The user-friendly handle is a unique identifier for the Target that is easy to remember and use in communication. The Identifier is a unique identifier that is used to differentiate the Target from others, while the Description provides more information about the Target's purpose and use.

The present state attribute indicates whether the Target is currently valid or invalid, which is important for ensuring that only valid data is used in business processes. The agency attribute specifies the role of the agency in managing the relationship between the Source and the Target. The allowable Validation Types attribute specifies the types of Sources that can be bridged to the Target. This helps to ensure that only valid and relevant Sources are linked to the Target.

The role that pays for related verifications involving the Target is also a confidential attribute. This specifies which role is responsible for paying for any verifications that are related to the Target. The list of data partners is a voluntary attribute that specifies which companies or organizations the Authority may share or sell data to. This helps to improve the Authority's compliance with privacy regulations and decrease risk and liability.

In addition to these confidential attributes, Targets also have public attributes that are discernible from the public record. These attributes include the Target's image, a CID hash that points to an immutable public business object, their Series identifier, and their Serial Number. These attributes are important for providing additional information about the Target and ensuring that the Target can be easily identified and distinguished from others.

The CID hash is a unique identifier that points to an immutable public business object that provides additional information about the Target. The Series identifier is a unique identifier that is used to group similar Targets together, while the Serial Number is a unique identifier that distinguishes one Target from another within a series.

Overall, the additional attributes generated by Targets are important for maintaining the integrity and validity of the Target, as well as providing additional information about its purpose and use.

12. Like Sources, there's no way to suss out a Target's relationships or even that it is a Target from publicly available data because all particles share the exact same public attribute footprint.

Targets, like Sources, share the same public attribute footprint and cannot be distinguished from one another based on publicly available data. This means that it is not possible to identify a Target's relationships or even that it is a Target without access to additional information that is kept private. The lack of differentiating data is intentional and is meant to protect the privacy and security of the information being shared through the Bridges.

The use of unique identifiers and cryptographic hashes in the creation of Targets ensures that they are easily distinguishable and verifiable by authorized parties. However, this information is not publicly available and is only accessible to those with the proper authorization and authentication.

This level of privacy and security is essential for businesses to protect their valuable information and prevent unauthorized access or breaches. By keeping the details of Targets private, businesses can maintain control over who has access to the information and ensure that it is only used for its intended purposes.

Overall, the ability to maintain the confidentiality of Targets and their relationships is crucial for businesses to maintain trust with their customers and partners. While it may limit the amount of publicly available information about Targets, this tradeoff is necessary to protect sensitive and valuable data.

13. Target attributes are immutable.

Targets, like Sources, have immutable and permanent records of their attributes. This means that any changes made to a Target's attributes will be permanently recorded and can be verified at a later time. This is crucial for reproof, verification, and legal nonrepudiation.

Reproof is the process of verifying that a particular action was taken or that certain data is accurate. For example, a business may need to reproof that a particular customer made a purchase on a specific date. With immutable records of Target attributes, businesses can easily verify this information and ensure that it is accurate and trustworthy.

Verification is the process of ensuring that a particular action or transaction meets specific criteria. For example, a business may need to verify that a particular product was produced according to certain quality standards. Immutable records of Target attributes can be used to verify that the necessary steps were taken to meet these standards and that the product is of high quality.

Legal nonrepudiation is the ability to prove that a particular action or transaction took place and that the parties involved cannot deny it. Immutable records of Target attributes can be used as evidence in legal proceedings to prove that a particular transaction occurred and that the parties involved cannot deny it.

Overall, the immutable record of Target attributes provides businesses with a reliable and trustworthy source of information that can be used for various purposes, including reproof, verification, and legal nonrepudiation. By ensuring that these records cannot be altered nor tampered with, businesses can have confidence in the accuracy and reliability of their data.

14. Verifications are never made starting with a Target. A verification is made from an Origin.

Verifications are always made starting with an Origin, not a Target. When a verification is requested, the Origin is the particle that initiates the process and provides the information to be verified. The Target serves as the endpoint for the verification but does not initiate or provide any information for the verification process. This is an important distinction to make, as it emphasizes the role of the Origin in the verification process and the importance of validating the information provided by the Origin before it is used for any downstream processes.

For example, let's say a Target is set up to allow access to help only from validated email addresses, and a business wants to verify the email the access of its customers. The business would initiate the verification process by submitting a list of email addresses to be verified. These email addresses would be the Origins in the verification process, and the Target (the access to help) would serve as the endpoint for the verification results. The verification process would validate the provided email addresses could indeed access the protected service. Meanwhile, the user might use the resulting Binary Link token which represents that their email was validated and that it was bridged to give them access to help, can now be bridged to another token that requires all of that previous chain – like access to chat with a support counselor. The Target particle would only be involved in the verification process if it's initiated either by an Origin upstream of it, or by the Binary Link token downstream of it.

This illustrates the importance of ensuring the validity and accuracy of Origins before they are used in downstream processes. If the Origins are not properly validated, the resulting Binary Link tokens and any subsequent processes based on them could be compromised or inaccurate. By emphasizing the role of the Origin in the verification process, businesses can ensure the integrity of their data and make more informed decisions based on accurate information.

15. Like Sources, Targets are unique unto themselves and there cannot be a copy of them.

Targets, like Sources, are unique and singular unto themselves. Once a Target is created, it cannot be duplicated or replicated. This is because each Target has a unique user-friendly handle and identifier that distinguishes it from all other Targets. The uniqueness of Targets is essential for maintaining the integrity and validity of the bridging relationships between Origins and Targets.

Furthermore, the singularity of Targets ensures that any verifications or audits related to a particular Target can be traced back to the original Origin that was bridged to it. The immutability of the record of their attributes also ensures that there is a permanent and tamper-proof record of the relationship between the Origin and the Target, as well as any relevant attributes.

The singularity of Targets is also important for compliance with regulatory and legal requirements. For example, in industries such as finance or healthcare, it is essential to maintain accurate and complete records of all transactions and relationships between parties. The uniqueness of Targets ensures that there is no confusion or ambiguity around the parties involved in any transaction or relationship, and that all relevant data can be easily traced back to the original Origin and Target.

In addition, the singularity of Targets also helps to prevent fraud and other malicious activities. Because each Target is unique and cannot be duplicated, it is much more difficult for bad actors to create fake or counterfeit Targets in an attempt to manipulate or deceive others. This is especially important in industries such as supply chain management or product authentication, where the validity and authenticity of a Target can have significant consequences for the entire chain of parties involved.

Overall, the singularity of Targets is an essential feature of the QBD protocol, as it helps to ensure the integrity and validity of bridging relationships between Origins and Targets, maintain compliance with regulatory and legal requirements, and prevent fraud and other malicious activities.

16. Targets are not possessable.

Targets are unique particles that serve as goals or desired outcomes for a business. Unlike Sources, Targets are not possessable, and they must remain in the treasury of the Governor of the platform. The Governor is the role responsible for ensuring the integrity, security, and management of the platform, and it will be covered in a later section. In this case, we can consider Data Stewardship Solutions as the Governor of a platform.

Targets cannot be possessed because they serve as endpoints for bridging relationships between Origins and Binary Links. Their unique attributes are immutable and permanent, and they are good for verification and nonrepudiation. By keeping Targets in the treasuring of the Governor, the platform can guarantee the integrity of the Bi-link Fabric created by all businesses.

The Governor can ensure the protection of Targets from unauthorized access or tampering, which is critical for maintaining the nonrepudiation of the platform. Nonrepudiation ensures that once an Origin has been validated and bridged to a Target, the relationship between the two cannot be disputed. Additionally, keeping Targets in the Governor's treasury also guarantees that the platform can maintain attestation, a critical component of verification, and auditing.

Attestation ensures that the platform can confirm the accuracy and integrity of a particular Target or Binary Link at any given time. It also allows for the creation of verifiable records, which can be used for auditing, compliance, or legal purposes. Therefore, by keeping Targets in the Governor's treasury, the platform can provide an additional layer of security and trust for all parties involved in a bridging relationship.

In summary, Targets are unique particles that serve as endpoints for bridging relationships between Origins and Binary Links. Unlike Sources, they are not possessable, and they must remain in the treasury of the Governor of the platform. This ensures the integrity and nonrepudiation of the Bi-link Fabric created by all businesses and guarantees that the platform can provide an additional layer of security and trust for all parties involved in a bridging relationship.

17. Targets do not have types.

Targets do not have a specific type, unlike Sources which can be assigned to a specific type. However, a Target can specify the types of Sources that can be bridged to them. This is done through a multi-select list of all possible Validation Types. The Authority who creates the Target can choose to limit the permissible Sources to specific types or allow any type of Source to be bridged. For example, a Target for a medical device company may only allow validated medical devices as Sources, while a Target for a certification program may only allow validated certifications as Sources. This ensures that the bridged relationships are valid and relevant to the Target's goal.

In addition, there is another type of Origin that can be bridged to Targets, known as the Binary Link. This specialized type applies only to Binary Link particles, which we will explore in greater detail in a future chapter of this methodology.

Having the ability to specify allowable Validation Types for a Target is important for maintaining the integrity of the bridged relationships. It helps to prevent irrelevant or invalid Sources from being bridged to a Target, which can create unnecessary risk and liability for the business. By limiting permissible Sources to specific types, the Target can ensure that the bridged relationships are relevant to its goal and that any verification or auditing can be traced back to the original Source.

Additionally, allowing for a multi-select list of all possible Validation Types provides flexibility for the Authority creating the Target. They can choose to limit permissible Sources to a specific type, or they can allow for a broader range of Validation Types to be bridged. This can be useful in situations where a Target may have multiple goals or use cases that require different types of Sources.

In summary, while Targets do not have a specific type assigned to them, they can specify the types of Sources that can be bridged to them. This is important for maintaining the integrity of the bridged relationships and ensuring that any verification or auditing can be traced back to the original Source. The flexibility of allowing for a multi-select list of all possible Validation Types provides additional options for the Authority creating the Target.

18. The Target Identifier must be unique.

The unique Target Identifier is a critical feature of the Target particle that ensures the integrity and reliability of the entire Bi-link Fabric. Without a unique identifier, it would be impossible to differentiate between different Targets, which could lead to confusion, errors, and potential breaches of security.

The Target Identifier is a user-friendly handle that is composed of a set of 3 characters (uppercase letters or numbers), followed by a hyphen, and then followed by four more such characters (e.g., BR5-49ME) that is similar to old phone numbers. Each character can be alphanumeric, which provides a vast range of possible combinations. This makes it easy for users to identify and work with different Targets and ensures that the unique identifier is easily memorable and recognizable.

The importance of the unique Target Identifier cannot be overstated. It is essential for businesses to have a reliable and robust system for identifying and managing Targets, especially in industries where compliance and regulatory requirements are strict. A unique identifier ensures that all Targets can be easily tracked and traced, which is essential for auditing, compliance, and legal purposes.

Furthermore, the unique identifier is immutable and permanent, which means that it cannot be changed or altered once it has been assigned. This provides a level of certainty and reliability that is critical in many industries, particularly those that deal with sensitive or confidential information.

In addition to providing a reliable way to identify Targets, the unique identifier also plays a crucial role in the verification and auditing processes. Because the identifier is immutable, it provides an unambiguous reference point for all verifications and audits that are performed on the Target. This ensures that all information related to the Target can be traced back to its source, which is essential for maintaining the integrity of the Bi-link Fabric.

Overall, the unique Target Identifier is a critical feature of the Target particle that ensures the reliability and integrity of the entire Bi-link Fabric. It provides a robust and reliable way to identify and track Targets, which is essential for compliance, regulatory, and legal purposes. By ensuring that the identifier is immutable and permanent, businesses can have confidence in the reliability and security of their data, which is essential for success in today's highly competitive and fast-paced business environment.

19. Targets automatically allow an Origin to be bridged to it that is of the type, "Binary Link".

In the coming years, it's expected that as AI continues to improve QBD, that eventually Binary Links will start to be informed as to their types through automation. But for revision 1.0, the type is simply Binary Link, and Targets must allow Bridges from them as a default behavior to allow for scalability. At the time of this writing, the architects were considering the ability for Targets to optionally elect to preclude Bridges from Binary Links.

Targets in Summary

Now that we've explored the fascinating world of Targets, we can see how these particles are the driving force behind businesses of all shapes and sizes. From quality control to project milestones to inter-industry sales, Targets are the relatable desires that keep organizations moving forward. But be warned, not identifying Targets can create risks and liabilities, so it's essential to define them well to unlock new opportunities and grow value inside an organization.

Just imagine an insurance company covering ambulances as they transport patients. What if some medical products traveling with the patient, such as a stent, had a Source token as it has been validated for use in that patient? By setting up a Target for validated procedure-ready medical devices, the insurance company could expand its business and take on a new revenue stream. And this is just one example of how Targets can lead to new business opportunities and growth.

However, it's crucial to remember that Targets must be created with care and consideration. Failure to identify and define Targets can lead to risks and liabilities, whereas well-defined Targets can unlock new opportunities and drive business growth.

So, if you're ready to harness the power of Targets and take your business to the next level, it's time to get started. Remember, Targets are created by someone with an Authority role by combining six attributes, including an Identifier, a Description, Agency, Allowable Validation Types, What Role pays for Verification, and a voluntary list of data partners. And with more attributes generated when Targets are created, the possibilities for business growth are endless.

What is Target Social Networking Communication (SNC)?

In our exploration of the properties of Targets within the Quantum Business Dynamics™ (QBD™) system, we now introduce a novel and exciting aspect—Social Networking Communication (SNC). This feature is designed to enhance the transparency, engagement, and interactive possibilities associated with each Target.

SNC is a property that can be assigned to a Target at the time of its creation by the Authority. The SNC attribute is a Boolean attribute—defaulting to false—that, when set to true, flags the Target's resultant Binary Links for potential public distribution and social engagement.

The inclusion of the SNC property introduces a new dynamic to the concept of a Target. In its default state (SNC = false), a Target acts as a point of interaction, linking Origins with the resultant Binary Links in a standard way. However, when SNC is activated (SNC = true), the Targets take on an additional role. Not only do they form the core interaction points within the QBD system, but they also become nodes for potential social networking communication.

The process operates as follows:

1. When a Target is created, the Authority has the option to set the makeSocial attribute to true, thereby making the Target SNC-activated. This decision is made based on the nature of the Target and the desired level of public engagement.

2. Binary Links generated from this Target, will have a heritage that includes the SNC attribute. For example, an initial aggregation of all resulting Binary Links from SNC-activated Targets can be quickly grouped. The SNC is used in this way to earmark a Binary Link as a potential point of interest for social communication.

3. A subscribed third-party, such as Data Stewardship Solutions (DSS), with proper authorization can monitor the creation of Binary Links associated with a specific set of SNC-activated Targets. After verifying permissions with the CNS for each Binary Link in the group, previously confidential information of these Binary Links can be distributed across subscribed social channels.
4. The social distribution process respects the confidentiality and permissions associated with each Binary Link. Only data that has been expressly permitted by the creator, or otherwise supported by applicable regulations for specified jurisdictions for social use is made available. This process ensures transparency, consent, and control over the data's use.

By enabling the SNC attribute on a Target, Authorities have the potential to transform a simple interaction into a public, socially engaging event. This addition enhances the utility of the Target, extending its functionality from being a purely internal component of the QBD system to a bridge between the system and the wider social ecosystem.

While the full benefits and possibilities of the SNC property will be explored in a later section, it is important to understand at this point that the SNC property has the power to add a dynamic and public-facing dimension to the concept of a Target within the QBD system.

Understanding Targets: Determination, Specification, and Vectorization

In the realm of Quantum Business Dynamics™ (QBD™), we often characterize the connections between Origins and Targets as vectors. These vectors represent unique pathways that different Origins can traverse to arrive at the same Target. But these pathways don't simply link point A to point B. As they embody the agreement between two entities, they also shape the nature of the possessable value that results from this agreement. This transformation is guided by predictable behavioral rules set by the Authority, resulting in a diverse range of Binary Links—much like the variations observed in DNA strands during mitosis.

Let's explore the unique properties of these "Targets"—Determination and Specification:

Determination: This property governs whether the resulting Binary Link from the Origin to the Target will be strongly or weakly typed. If the determination is set to "undetermined" during the Target's creation, the ensuing Binary Link's Validation Type will be untyped and carry a reserved integer that represents this "undetermined" state. Drawing a parallel with biology, these Binary Links can be compared to stem cells due to their potential to connect with any type of Target.

Specification: This property outlines the logic for determining the Binary Link's Validation Type based on its Origin. If the Target's determination switches from "undetermined" to "determined," the Authority (the Role managing the Target) has the option to leave the resulting Validation Type "unspecified." In this scenario, the Binary Link type will automatically inherit the Origin's Validation Type, culminating in a strongly typed link.

However, if the Authority decides to specify a unique determination pattern for each type of Origin, it can prompt a "mutation" in the Binary Link Validation Type. For instance, an Origin of 'email' Validation Type could result in a Binary Link of 'website ID' Validation Type.

Summarizing, we end up with three primary scenarios:

- **Undetermined (U)**: The resulting Binary Link acts like a "Stem Cell," capable of connecting to any type of Target.

- **Determined-Unspecified (DU)**: The resulting Binary Link inherits its Validation Type from the Origin, akin to genetic inheritance.

- **Determined-Specified (DS)**: The resulting Binary Link undergoes a "mutation," with its Validation Type determined by a unique rule set by the Authority for each resulting of Origin.

The strategy of vectorizing Targets provides a potent mechanism to model intricate relationships and transformations in a system, mirroring the way different nucleotide sequences encode diverse traits in DNA. By understanding and applying these principles, executives can better chart growth opportunities as they establish Targets that will attract various types of Origins. Meanwhile, data scientists and business analysts can harness the power of QBD to make more informed decisions and predictions.

Now that we've covered Targets in-depth, it's time to move onto the next step in the methodology: Bridges. So, buckle up and get ready to dive into the world of Bridges, but first, by way of transition, here are a few words about how Bridge Durability relates to Targets.

Introducing Bridge Durability

In Quantum Binary Dynamics™ (QBD™), understanding the structure and dynamics of Targets and Bridges is fundamental. As you've learned about Targets, you've also gotten glimpses of Bridges and their role in this intricate system. Now, let's expand on an important property of a Target that directly affects the Bridges resulting from it—Bridge Durability.

In essence, the Bridge Durability property of a Target is the mandate given by the Authority on how long-lasting or enduring the Bridges associated with the Target can be. This property essentially sets the rules for the lifetime of a Bridge—whether it is Timeless, implying that it will last conceivably forever without end (unless invalidated by an authorized external action), or Temporal, meaning it has a pre-determined lifespan.

1. **Timeless**: The default setting for Bridge Durability. If the Authority chooses Timeless when casting a Target, the Bridges created from this Target are designed to endure indefinitely. They exist in perpetuity until they're invalidated by a specific action from an authorized entity.

2. **Temporal**: If the Authority sets Bridge Durability to Temporal when casting a Target, all resulting Bridges must comply. They are inherently designed with an expiration, a lifespan after which they automatically become invalid. This lifespan is defined by a number (count) and an increment (like seconds, days, eons)—for instance, 2 years. At this level, the Authority specifies the lifespan for all resulting Bridges that represent an agreement between two parties regarding a selected Origin and this Target.

Note: If a Target is set to Timeless, a Broker creating a Bridge has the option to alter the Bridge Durability to Temporal. Conversely, if a Target's Bridge Durability is Temporal, this cannot be overridden by the Broker to Timeless.

In a nutshell, the Bridge Durability property of a Target provides granular control over the lifespan of Bridges. This powerful feature offers versatility in creating Bridges that meet various needs, from short-term projects to long-term initiatives and everything in between. We'll delve deeper into the workings and implications of Bridge Durability in the coming sections. For now, this understanding sets a solid foundation for your further exploration of QBD's robust system.

Introducing Bridge Venue Selection

Prepare to be amazed as we unveil another groundbreaking aspect of

Quantum Binary Dynamics™ (QBD™)—Bridge Venue Selection. In the realm of legalities and jurisdictional considerations, QBD empowers you with unprecedented control over the resolution of potential disputes related to Bridges brokered to your specific Targets. With Bridge Venue Selection, you hold the power to designate the specific jurisdictions where legal actions, if ever needed, should take place.

Imagine having the peace of mind knowing that in case of any disagreements or conflicts arising from a Bridge brokered to one of your Targets, the venue for dispute resolution has already been predetermined, providing clarity and certainty for all parties involved. The genius lies in the flexibility it offers, granting you the ability to strategically choose venues that align with your business objectives, regulatory requirements, and preferred legal environment.

As an Authority setting up a Target, you hold the reins of power, guiding the creation of Targets that align with your strategic vision. Among the various settings available, Bridge Venue Selection introduces an optional advanced feature that can significantly influence the resolution of potential disputes related to the Bridges brokered to your Targets.

When you begin crafting a Target, by default, the Bridge Venue is set to "Venueless," meaning that Bridges can operate without constraint for jurisdictional considerations governing any potential disputes. In this scenario, any legal actions or disputes arising from the Bridges will be subject to the existing laws of the parties involved.

However, if you wish to add an extra layer of clarity and control, you can easily set the acceptable Bridge Venue(s) for your Target. By selecting one or multiple venues from the diverse array provided by QBD, you can align the jurisdiction for any subsequent Bridges brokered to your Target with your business objectives, regulatory preferences, and legal comfort.

Imagine this process as setting the geographical boundaries that can greatly limit your risk for adopting the power of QBD. Your Bridge Venue selection acts as a friendly guide that provides choices to Brokers and adds clarity for any potential legal actions to use a specific jurisdiction, providing peace of mind and certainty for all parties involved.

Notably, as you create your Target, the Bridge Venue Selection appears as a natural part of the process. A thoughtful decision in this advanced settings empowers you to proactively shape the jurisdictional considerations of your Target, even if you choose not to specify a venue. Remember, your Target's settings are a reflection of your strategic vision for your business, and QBD offers the flexibility to adapt these choices as your needs evolve.

For Brokers who bridge to your Target, if you haven't specified a Bridge Venue, they can freely create Bridges without any predefined jurisdictional constraints—and they cannot further constrain venue. However, should you decide to set a specific Bridge Venue, Brokers will be required to choose a pure subset of these venues for their Bridges, which could be all of them. This elegant mechanism ensures alignment and compliance with your Target's jurisdictional preferences.

In this realm of legalities and jurisdictional considerations, QBD empowers you with unprecedented control over the resolution of potential disputes related to Bridges brokered to your Targets. Setting the Bridge Venue is a testament to your strategic vision and your commitment to building a robust and transparent business ecosystem within QBD.

As you explore the vast possibilities of Quantum Business Dynamics™, Bridge Venue Selection presents a powerful opportunity to make a profound impact on your business landscape. Embrace this exciting dimension and unlock the true potential of your Targets in the dynamic world of QBD. Let the journey begin!

A Word about Bridge Venue and Freedom

In the ever-evolving landscape of global business and digital transformation, Quantum Binary Dynamics™ (QBD™) offers an unprecedented opportunity to reshape the way we approach disputes and legal resolutions. As we the ripple effects from using Bridge Venue Selection, it's essential to recognize the profound impact this optional advanced feature can have on promoting essential freedoms, data privacy rights, and fostering a transparent business environment.

As you have read, at the core of Bridge Venue Selection is the power to designate specific jurisdictions where legal actions related to Bridges brokered to your Targets should take place. This seemingly simple yet influential setting allows you, as an Authority, to proactively align your business objectives with regulatory requirements and preferred legal environments. By strategically choosing venues, you can add an extra layer of clarity and control to your Targets, ensuring that any potential disputes are resolved according to your vision.

However, beyond its pragmatic implications, Bridge Venue Selection also serves as a powerful catalyst for promoting positive social change. As QBD continues to gain global adoption, the Bridge Venue choices made available through the platform carry significant weight, encouraging governments, businesses, and institutions to consider the fundamental concepts of ownership, data privacy rights, and essential freedoms.

In this interconnected digital era, the benefits of QBD extend beyond economic growth and efficiency. By aligning Bridge Venues with jurisdictions that recognize and respect these foundational rights, QBD fosters a business ecosystem that thrives on transparency, trust, and ethical principles. As businesses and governments embrace the power of QBD, they naturally gravitate towards venues that support inalienable rights, thereby contributing to the wellbeing and harmony of societies.

While QBD does not seek to enforce laws or regulations, the choices made by Authorities and Brokers within the platform can exert positive pressure for change. As more regions and states witness the transformative potential of QBD and the associated Bridge Venues, there emerges a tangible incentive for adopting policies and legal frameworks that align with the principles upheld within the QBD ecosystem.

As Authorities and Brokers exercise their autonomy in choosing Bridge Venues that reflect their commitment to data privacy, ownership, and essential freedoms, the ripple effects are felt beyond the confines of QBD itself. These choices become a beacon of progress, influencing the evolution of laws and regulations in various jurisdictions.

By embracing Bridge Venue Selection and consciously selecting venues that promote fairness, transparency, and individual rights, QBD participants play an active role in shaping a more just and equitable global business landscape. As bridges are established and connections strengthened across borders, the Bridge Venues chosen serve as a testament to the shared values of the QBD community.

Furthermore, the flexibility offered by Bridge Venue Selection creates opportunities for businesses operating in regions with limited freedoms to seek alternative jurisdictions that better protect their interests. This fosters healthy competition among jurisdictions, encouraging governments to enhance their legal frameworks to attract and retain businesses within their borders.

In essence, QBD becomes more than just a platform for efficient business transactions—it becomes a catalyst for positive change. By promoting essential freedoms and data privacy rights, Bridge Venue Selection empowers businesses to thrive in an environment of trust and security. The symbiotic relationship between QBD and the jurisdictions it touches fosters a virtuous cycle of growth and progress.

As QBD continues to expand its global reach, the choices made within its framework contribute to the broader evolution of societies, undergirding a world where the principles of fairness, accountability, and ethical conduct reign supreme. The journey towards a more interconnected and socially responsible world begins with the choices made in Bridge Venue Selection—an opportunity for both businesses and governments to embrace a shared vision of progress and prosperity.

So, as you continue on your journey with Quantum Binary Dynamics™ and delve into the power of Bridges and their related agreements, remember that your choices have the potential to ignite positive change, leaving a lasting impact not only within QBD but across the vast expanse of the global business landscape. Embrace the power of Bridge Venue Selection, and together, let us shape a brighter future for businesses and societies worldwide.

Bridges

Bridges are the magic in the realm of Quantum Business Dynamics™! As a business executive, you are no doubt aware of the many challenges that your organization faces in the current rapidly changing and competitive landscape of the digital economy. With increasing amounts of data being generated every second, there is an urgent need for businesses to create systems and processes that can keep up with this explosion of information.

That's where Bridges come in. Bridges are an essential part of the Quantum Business Design methodology, and they offer a way for businesses to create immutable links between two entities, providing a permanent record of their relationship. With Bridges, businesses can create an infrastructure that is transparent, secure, and traceable.

One of the most exciting features of Bridges is their immutability. Once a Bridge is created, it cannot be updated nor modified. This means that businesses can rely on the information contained within the Bridge to be accurate and valid, making it an ideal tool for verification and auditing purposes. Additionally, Bridges offer a way for businesses to maintain the privacy and security of their data, ensuring that sensitive information is not accessible to unauthorized parties.

Another key feature of Bridges is their user-friendly handle. This unique identifier allows businesses to quickly and easily identify the relationship between two entities, making it easy to manage and maintain their data. With Bridges, businesses can rest assured that their data is secure, and that they can easily trace the history of any changes that have been made.

But Bridges aren't just about data security and privacy. They also play a crucial role in building trust between businesses and their customers. By assigning an Authorized Agent as the Rights Resource Manager for each Bridge, businesses can demonstrate their commitment to protecting their customers' information rights and data privacy. This is where ethics get built into the system, and where Quantum Business Dynamics™ provides Ethics as a Platform, removing risk for companies who adopt QBD.

Additionally, Bridges offer a way for businesses to comply with the complex and ever-changing regulations of different Jurisdictions. By providing a clear and traceable path between the Bridge and its predecessor Origin, as well as between the Bridge and its descendant Target, businesses can ensure that they are in compliance with the relevant regulations and laws.

Finally, Bridges are a tool for innovation and growth. By creating immutable links between entities, businesses can identify new opportunities for collaboration and partnership. With Bridges, businesses can quickly and easily establish new relationships and create innovative solutions to complex problems.

Lastly, Bridges offer an exciting and innovative solution for businesses looking to navigate the challenges of the digital economy. By providing a way to create secure and transparent links between entities, Bridges can help businesses maintain the privacy and security of their data, build trust with their customers, comply with regulations, and identify new opportunities for growth and innovation. We invite you to explore the exciting world of Bridges and see how they can help take your business to the next level.

1. Bridges must have a single predecessor and a single descendant.

The first feature of Bridges is that they must have a single predecessor and a single descendant. A Bridge always links two particles: the predecessor, which must be an Origin, and the descendant, which must be a Target. Origins can be either Sources or Binary Links, while the descendant must always be a Target.

This means that each Bridge serves as a crucial link between a validated Origin and a desired Target, helping to maintain the integrity of the entire QBD system. It also ensures that each Bridge is easy to trace back to its original Origin, making it easier to verify the validation and credibility of the particles involved.

Without this single predecessor and descendant rule, the system could become chaotic and difficult to manage. If Bridges could connect multiple Origins to a single Target or vice versa, this would lead to confusion about which Origins were valid and which were not. The single predecessor and descendant rule helps to avoid this confusion and keeps the system organized and efficient.

This feature is crucial to understanding how Bridges work within the larger QBD system. Each Bridge represents a relationship between a single Origin and a single Target, and it is this relationship that allows businesses to establish trust and credibility with their partners and customers. Without Bridges, the QBD system would not be able to function effectively or provide the level of transparency and validation that businesses need to succeed.

2. Bridges are created by someone with the role of Broker.

In QBD, Bridges are created by someone with the role of Broker. It is crucial that the Owner of the Origin is aware of the agreement and has some ability for affirmative express consent. Part of the genius of QDB architecture is that it supports an automatic way to achieve this in its native functionality. If the Origin is still in the Governor's treasury, then the Owner may not be aware of the Origin and any Bridge created for it will have the state of "Aspirational" and cannot be considered Valid for any subsequent Verification. However, if the Origin has left the treasury, there is effectively no technical way for the Broker to create a Bridge unless the Owner or their delegated Holder is part of the process, at which point, any subsequent Bridge would be automatically considered Valid.

The authorized Broker builds the Bridge as a result of an agreement between two people, one of whom is the Owner of the Origin and the other is the Authority for the Target, who provides their agreement as part of creating the Target. The Broker is responsible for ensuring that the Bridge is valid and can be used to connect the two particles. The Broker's role is crucial in maintaining the integrity of the QBD system and ensuring that all Bridges are valid.

Bridges are the result of intent derived from a process, product action, systems, or services that identify an agreement between two people that involve a Source and a Target. This intent is translated into a Bridge, which acts as a tangible representation of the agreement. This makes it easier for all parties involved to understand and manage the agreement, as well as providing a record for auditing and compliance purposes.

In the words of Mike Metz, a principal analyst at Gartner, "bridging the gap between the business and IT is always a challenge. But with the right approach, tools, and technology, it can be done." Bridges serve as that tool to span the gap between different parties involved in the agreement, whether they are from the same organization or from different organizations.

One example of how Bridges can be used is in the real estate industry, where agreements between buyers, sellers, and agents need to be recorded and tracked. Bridges can be used to connect the different parties involved in the agreement and to create a record of the agreement that can be used for verification and auditing purposes. This ensures that all parties involved in the agreement are on the same page and that there is a record of the agreement that can be referred to if needed.

3. Bridges can be part of a verification from an Origin to a Target.

Bridges play an important role in the verification process between a Source and a Target. When a Validator creates a binary relationship between a Source and a Target, they establish the agreed-upon criteria for the relationship. The Verification process then relies on this established relationship between the two particles to ensure that the facts presented are accurate and remain valid over time. Verifiers are responsible for reviewing these established binary relationships and their corresponding Binary Link tokens to independently verify the claims presented in the relationship.

The process of Verification saves time and money by not requiring the Verifier to start from scratch when reviewing a binary relationship. Instead, they can rely on the previously established criteria and relationship to arrive at a level of confidence required for their task. This helps to avoid costly and time-consuming re-validation and reduces the risk of conflicting sets of understanding. By verifying that the binary relationship between the Source and the Target is accurate and valid, Verifiers can help to ensure that the facts presented are reliable and trustworthy.

Bridges can also be involved in this process as the endpoint for the results of the verification. When a business initiates a verification process, the Origins involved in the process serve as the starting point, and the resulting Binary Link token serves as the verification record. The Target, which is the endpoint for the verification results, can be linked to the Binary Link token downstream of it. This helps to maintain the integrity and validity of the Bi-link Fabric and ensures that the verification record can be traced back to the original Origin.

So, Bridges are an integral part of the verification process in QBD. They establish the criteria for the binary relationship between a Source and a Target, which is then used in the verification process to ensure the accuracy and validity of the facts presented. Verifiers can rely on these established binary relationships to save time and money and reduce the risk of conflicting sets of understanding. By using Bridges in the verification process, businesses can ensure the reliability and trustworthiness of the facts presented and maintain the integrity of their Bi-link Fabric.

4. Bridges are immutable and must be replaced rather than updated.

The fourth feature of Bridges is that they are immutable and cannot be updated. If any changes need to be made to a Bridge, it must be replaced entirely with a new one. This may seem like an inconvenience at first, but it is actually an important aspect of maintaining the integrity and verifiability of the Bi-link Fabric.

By creating immutable Bridges, businesses can be sure that the relationships established between Sources and Targets remain accurate and reliable over time. If a Bridge could be updated, there would be no way to know whether the information contained within it was valid at any given point in time. Additionally, updating a Bridge could potentially create conflicts with other established Bridges, leading to confusion and errors in the verification process.

Replacing a Bridge entirely instead of updating it ensures that any changes made to the relationship between a Source and a Target are properly documented and can be audited for compliance or verification purposes. This helps to create a more transparent and trustworthy system for all parties involved.

While it may seem like a cumbersome process to replace Bridges rather than simply updating them, it actually saves time and resources in the long run. By avoiding the need to constantly update and validate Bridges, businesses can focus on other aspects of their operations and trust that the established Bi-link Fabric is accurate and reliable.

Overall, the immutability of Bridges is a key feature of QBD that helps to maintain the integrity of the Bi-link Fabric and establish trust between all parties involved. By creating a reliable and verifiable system for establishing relationships between Sources and Targets, businesses can reduce risk and liability and focus on achieving their goals with confidence.

5. Bridges are considered valid when they are created.

Bridges are considered valid at the moment of their creation. This means that when a Broker establishes a Bridge between an Origin and a Target, the agreement between those two parties is considered valid at that moment. This validity is reflected in the resulting Binary Link particle that gets generated and can be used for verification purposes.

It's important to note that this validity only applies at the moment of creation. Over time, the validity of a Bridge can be affected by changes in the underlying data or circumstances. For example, if a Target's requirements change, a previously valid Bridge may no longer be valid.

This is why it's important to establish a robust verification process that can ensure the ongoing validity of Bridges. The process of verification involves checking the accuracy and authenticity of the data underlying the Bridge, as well as verifying that the Bridge still meets the criteria for validity.

In some cases, it may be necessary to replace a Bridge rather than simply updating it. This is because Bridges are immutable, which means that they cannot be modified after they have been created. Instead, a new Bridge must be created to replace the existing one. This can be done by creating a new Binary Link particle that represents the updated agreement between the Origin and the Target, and then replacing the original Bridge with the new one.

By ensuring the ongoing validity of Bridges, businesses can maintain the integrity of their data and processes and avoid potential legal or financial risks. Valid Bridges can help to establish trust between parties and provide a basis for ongoing collaboration and partnership. Additionally, by ensuring that Bridges are valid and up-to-date, businesses can be more responsive to changes in the market or industry and adapt their processes and strategies accordingly.

In summary, Bridges are considered valid at the moment of creation, but this validity can be affected by changes over time. Establishing a robust verification process is essential for maintaining the ongoing validity of Bridges and avoiding potential risks. By ensuring the validity of Bridges, businesses can build trust with partners, respond more effectively to changes in the market, and maintain the integrity of their data and processes.

6. The state of a Bridge can be subscribed to.

The sixth feature of Bridges is that their state can be subscribed to, and any updates or superseding of the Bridge will notify all subscribers. This feature provides a mechanism for interested parties to stay informed about the state of the Bridges they care about. For example, if a company relies on a Bridge to conduct business with a particular partner, they can subscribe to the Bridge and be notified if any changes or updates are made to it. This helps to ensure that all parties are on the same page and that there are no misunderstandings or miscommunications.

The subscription feature also makes it easier to manage and maintain Bridges. Rather than having to manually inform all parties involved in a Bridge every time there is an update or change, subscribers can be automatically notified. This helps to reduce the risk of errors or oversights and ensures that everyone stays informed in a timely manner.

However, it's important to note that Bridges can only be subscribed to if they have been properly validated and established in the first place. If a Bridge is created without proper validation or is otherwise invalid, there will be no subscribers to notify of updates or changes. This underscores the importance of properly validating Bridges before they are created, and the need for ongoing monitoring and maintenance to ensure that Bridges remain valid over time.

Overall, the subscription feature of Bridges is a valuable tool for businesses and organizations that rely on them to conduct important transactions or relationships. By staying informed of updates and changes, parties can ensure that they are always on the same page and that there are no misunderstandings or miscommunications.

7. The agency of the Target is inferred as the agency of the Bridge.

The agency of a Target is a critical attribute, as it helps to establish the context in which the Target will be used. In the case of Bridges, the agency of the Target is also used to infer the agency of the Bridge. This is important because it helps to ensure that the Bridge is properly aligned with the overall goals and objectives of the agency that is responsible for managing the Target.

In addition to inferring the agency of the Bridge, an Agent from that genre of agency can optionally be assigned to manage the Bridge. This is an important feature because it helps to ensure that the Bridge is properly managed and that any issues or concerns are addressed in a timely and effective manner. By assigning an Agent to manage the Bridge, the agency can be confident that the Bridge is being properly monitored and that any necessary updates or changes are being made as needed.

Overall, the ability to assign an Agent to manage a Bridge is a powerful tool that can help to ensure the success of the Bridge and the overall goals of the agency. By properly managing Bridges, agencies can help to ensure that they are meeting their objectives and providing the best possible service to their customers and stakeholders.

8. The act of creating a Bridge between a Source and a Target is the catalyst for the combination to yield a Binary Link token downstream of the Target.

The eighth feature of Bridges is that the creation of a Bridge between a Source and a Target is the catalyst for the generation of a Binary Link token downstream of the Target. This is an essential component of the QBD methodology, as the Binary Link token is the record of the established relationship between the Source and the Target. When a Bridge is created, it serves as the connection between the Source and the Target, enabling them to be linked upstream and downstream.

This Binary Link token acts as an immutable and permanent record of the established relationship, and it is generated automatically as part of the creation of the Bridge. The Binary Link token contains important information about the relationship, including thc time and date the Bridge was created, the identities of the Source and Target, and any other relevant details.

The creation of the Binary Link token downstream of the Target is an important aspect of the QBD methodology. It ensures that the relationship between the Source and Target is fully traceable and auditable, allowing for easy verification of the validity of the relationship. The Binary Link token is also immutable and permanent, meaning that it cannot be altered nor deleted once it has been created, adding an additional layer of security and authenticity to the relationship.

The creation of a Bridge and the subsequent generation of a Binary Link token is a critical component of the QBD methodology. It ensures that the relationship between the Source and Target is fully documented and traceable, providing a high level of trust and transparency in the relationship. By automating the generation of the Binary Link token, QBD saves time and effort while also ensuring accuracy and reliability in the documentation of the relationship.

9. Bridges can only be linked upstream to Origins that have been bridged to them.

One important feature of Bridges is that they can only be linked upstream to Origins that have been bridged to them. This means that before a Bridge can be created, there must be a Binary Link established between the Source and the Bridge's predecessor Origin.

This feature ensures that there is a clear and traceable path between the Bridge and its predecessor Origin, as well as between the Bridge and its descendant Target. It also reinforces the idea that all Binary Links must be established and validated before a Bridge can be created, which is important for maintaining the integrity and security of the overall system.

By linking upstream to only previously validated Origins, Bridges ensure that only validated and trusted data is used in the creation of the Binary Link. This reduces the risk of errors or fraudulent activity in the system and increases the overall trustworthiness of the data being used.

Additionally, this feature ensures that Bridges can be easily traced back to their source, allowing for easy verification of the Binary Links established by the system. This is important for maintaining compliance with regulations and for ensuring that the system remains trustworthy and reliable.

10. Bridges must have a user-friendly handle.

The handle of a Bridge is a user-friendly identifier that is assigned to the binary relationship between a Source and a Target. This identifier is a unique set of 3 characters (uppercase letters or numbers), followed by a hyphen, and then followed by four more such characters (e.g., BR5-49ME) similar to old phone numbers. Unlike Sources and Targets, which must have singularly unique handles, Bridges can be superseded by new Bridges to allow for changing conditions. However, the handle of a Bridge must be unique at the time of creation, differentiating it from all other Bridges.

The handle of a Bridge plays a crucial role in the identification and tracking of binary relationships within the QBD methodology. The use of a user-friendly identifier makes it easy for people to remember and access information about specific binary relationships. Moreover, this identifier can be used in conjunction with other metadata to provide a complete picture of a given binary relationship.

In addition to its role as a unique identifier, the handle of a Bridge is also inherited by all superseding Bridges that represent updated states for the agreement between two parties. This inheritance ensures that all Bridges related to a particular binary relationship are easily identifiable and traceable. It also ensures that any downstream Binary Links that are established based on a particular Bridge are associated with the correct binary relationship.

Overall, the user-friendly handle of a Bridge serves an important function within the QBD methodology. It ensures that binary relationships are easily identifiable and traceable, even as they evolve and change over time. By assigning a unique handle to each Bridge, the QBD methodology provides a powerful tool for managing and tracking binary relationships within a business or organization.

11. A Rights Resource Manager (_RRM) is assigned to a Bridge when it is created.

One of the key features of Bridges is that a Rights Resource Manager (_RRM) is assigned to them when they are created. The _RRM is selected from a list by the Validator and is typically an Authorized Agent (AA). An AA is a special kind of nonprofit organization that is certified to represent the information rights of the user for whom the agreement originated. The user is considered the Actor on the system whose actions are creating data that will later be monetized.

There are many good AAs to choose from, such as the Privacy Co-op, the Data Dividend Project, and Consumer Reports' Permission Slip. By assigning an AA as the _RRM from the very start, QBD accomplishes the important goal of safeguarding users' information rights and data privacy. This is where ethics get built into the system, and where QBD provides Ethics as a Platform (EaaP).

By using an AA as the _RRM, Bridges provide an important component of the ability to verify any data use against any combination of jurisdictions' regulations for the related Source, which facilitates compliance. This is extremely important as it removes risk for companies who adopt QBD. With the increasing importance of data privacy regulations around the world, companies need to take extra care to ensure that they are complying with all applicable regulations.

The _RRM is responsible for managing the rights of the user in relation to the data being generated by the Bridge. This includes things like ensuring that the user's information is being used in accordance with their wishes, and that their data is being stored securely. The _RRM also can enforce the user's rights in the event of a violation. This means that if a company violates the user's data privacy rights, the _RRM can take action to protect the user and enforce their rights.

Overall, the assignment of an AA as the _RRM is an important feature of Bridges, as it ensures that users' data privacy rights are being protected and that companies are complying with all applicable regulations. This is an essential component of QBD, as it removes risk for companies and helps to build trust with users.

12. When created, Bridges naturally generate new attributes beyond the original three that caused their creation.

When a Bridge is created, it generates new attributes beyond the original three that led to its creation. These attributes are divided into two categories: confidential and public. Confidential attributes are those that are kept private unless there is a need and authorization to share them. The confidential attributes of a Bridge include its user-friendly handle and its current state, which can be either valid or invalid.

On the other hand, public attributes are those that are discernible from the public record. These attributes are designed to be easily accessible to anyone who has permission to access the Bridge. The public attributes of a Bridge include a CID hash that points to an immutable public business object, its Series identifier, and its Serial Number.

It's important to note that no other data needs to be disclosed beyond these public attributes. This provides a layer of protection for the information contained within the Bridge, while still allowing for easy access to important information for authorized parties.

The inclusion of these attributes is a key component of the QBD methodology, as it allows for transparency and accountability in the management of Bridges. By keeping the confidential attributes hidden and the public attributes easily accessible, users can have confidence that their information is being handled with care and that their rights are being respected.

Overall, the inclusion of both confidential and public attributes in the creation of Bridges helps to ensure that data is being handled responsibly and ethically. It allows for transparency and accountability while also providing the necessary privacy and protection for sensitive information. By adhering to these principles, companies can use QBD to effectively manage their data and minimize risk.

13. Like Sources and Targets, there's no way to suss out a Bridge's relationships.

The 13th feature of Bridges is their anonymity. Like Sources and Targets, Bridges also lack any identifying markers in their public attribute footprint, making it impossible to differentiate them from other particles or determine their function. This is a key security feature of the QBD system, as it ensures that only those with the proper authority can access the sensitive information contained in the Bridges.

Without the ability to differentiate Bridges from other particles, bad actors are unable to identify and target them for nefarious purposes. This creates an added layer of security and protects sensitive information from falling into the wrong hands. Additionally, it provides a sense of privacy to those involved in the agreement represented by the Bridge, as their personal information is not easily accessible.

However, this anonymity also presents a challenge for those tasked with managing and monitoring the Bridges. In order to keep track of them,

Validators must rely on their own internal record-keeping systems, as well as the system's audit trail, to keep track of the Bridges' creation and changes over time.

The anonymity of Bridges is also important for compliance purposes. Because they cannot be easily identified from their public attributes, they can only be accessed by those with the proper authority and permissions. This allows for strict adherence to regulations around data privacy and security, as access to sensitive information is strictly controlled.

Overall, the anonymity of Bridges is a critical component of the QBD system's security and privacy measures. While it presents a challenge for those managing and monitoring the Bridges, it ultimately ensures that sensitive information is kept safe and secure and provides peace of mind to those involved in the agreement represented by the Bridge.

14. Like Sources and Targets, the record of their attributes is immutable and permanent, good for reproof, verification, and even legal nonrepudiation.

Bridges, like Sources and Targets, have an immutable and permanent record of their attributes. This feature is crucial because it allows for easy reproducibility, verification, and even legal nonrepudiation of agreements between two parties. Immutable records provide a guarantee that the original data cannot be tampered with, which is important for building trust between parties in any agreement.

Moreover, immutable records can help with regulatory compliance and legal disputes. The permanent record of Bridge attributes ensures that any agreements made between parties can be traced and verified, allowing for easier compliance with regulatory requirements. Additionally, in case of legal disputes, the immutable record of Bridge attributes can serve as evidence, providing a permanent and trustworthy record of the agreement made between two parties.

This feature is also important for business operations. Immutable records can help with internal auditing, ensuring that agreements made between parties are adhered to and that the business is operating in a transparent and trustworthy manner. Additionally, it can help with external auditing and due diligence, providing an easy and trustworthy source of information for auditors and potential investors.

Overall, the immutable and permanent record of Bridge attributes is a crucial feature that provides a high level of trust, transparency, and reliability in any agreement made between two parties. It is an essential component of the QBD methodology, providing a strong foundation for the secure and efficient exchange of data between parties.

15. Bridges are immutable and therefore will be superseded as changes are needed.

When a Bridge needs to reflect changes to the agreement it represents, such as when it is invalidated or when the user wants to change their Agent or _RRM, a new Bridge is created to supersede the old one. This is because Bridges are immutable and cannot be updated directly. Instead, creating a new Bridge with updated information maintains the integrity of the record while also allowing for necessary changes.

The new Bridge will inherit the handle of the previous Bridge, which is already unique in describing the relationship between a specific Origin and Target, ensuring that it can be traced back to the original agreement. The old Bridge will still exist in the system and will remain immutable, providing a permanent record of the original agreement.

When a superseding Bridge invalidates a previously valid bridge, no new Binary Link token is yielded. However, when a superseding Bridge results in a reversal of an invalid state, the new valid Bridge will yield a new Binary Link on the downstream side of the Target. This results in two Binary Link Tokens that represent the same agreement relationship between the same Origin and Target. The first token is reliable for verification of the fixed time period from the point of the original agreement until it was subsequently invalidated. The new token is then reliable for verification of the time period from its reinstatement as valid until the present. It's important to note that both Binary Link Tokens can be verified over and over again, allowing for a comprehensive and immutable record of the entire agreement relationship.

This feature allows for the evolution of agreements over time while maintaining a clear and traceable history. It also provides a means of resolving disputes that may arise by providing a permanent record that can be used for reproof or legal nonrepudiation. Overall, the superseding of Bridges provides a powerful tool for managing agreements and ensuring their accuracy and integrity.

16. All Bridges must contain information about the Bridge they replace.

Bridges are unique particles that serve as the link between a Source and a Target. They are immutable and must be replaced rather than updated, which means that when one needs to change to reflect changes to the agreement between two humans that it represents, they are superseded by a new Bridge. One interesting feature of Bridges is that the new Bridge inherits the handle of the previous Bridge, which is already unique in describing the relationship between a specific Origin and Target, ensuring that it can be traced back to the original agreement. The old Bridge will still exist in the system and will remain immutable, providing a permanent record of the original agreement.

Another feature of Bridges is that all Bridges must contain information about the Bridge they replace. This is important for maintaining the integrity of the system and ensuring that the history of the agreement is preserved. When a new Bridge is created to replace an old one, the old Bridge will still exist in the system, but it will contain dummy data if it was the prime Bridge. This means that no one can discern a difference between the old and new Bridges, and therefore it remains impossible to identify patterns or relationships without appropriate authorization to do so.

The requirement that Bridges contain information about the Bridge they replace ensures that the history of the agreement is preserved, and that the system can maintain the integrity of the Bridges. This feature also contributes to the overall security of the system, as it allows authorized individuals to verify the history of the agreement and ensure that there have been no unauthorized changes to the Bridges. Additionally, this feature helps to prevent fraudulent activity by making it more difficult for individuals to manipulate or falsify Bridges in order to gain access to sensitive information.

Again, Bridges are an important component of the QBD methodology, serving as the link between a Source and a Target. They are immutable and must be replaced rather than updated and contain information about the Bridge they replace to ensure the history of the agreement is preserved. This feature is essential for maintaining the integrity of the system and preventing fraudulent activity, contributing to the overall security of the QBD methodology.

17. Bridges are not possessable.

The concept of Bridges not being possessable means that they are not owned by any single entity, whether it is a person or a company. Instead, they represent a shared understanding between two parties, and as such, they cannot be bought, sold, nor transferred.

This is an important feature of QBD because it helps to promote transparency and trust in the system. By ensuring that Bridges are not possessable, it eliminates any conflicts of interest that may arise if one party were to control the Bridge. This means that the agreement between two parties is based on mutual understanding and trust rather than a hierarchy of power.

Additionally, the fact that Bridges are not possessable means that they are not subject to the same legal restrictions and regulations as other types of assets. This can help to simplify the legal framework around QBD and make it easier for companies to adopt the technology.

However, it's important to note that while Bridges themselves are not possessable, the information they represent may still be subject to legal ownership and privacy restrictions. For example, a Bridge may represent a contract between two parties, but the information contained in that contract may be subject to copyright or other intellectual property laws.

Overall, the fact that Bridges are not possessable is a key feature of QBD that helps to promote transparency, trust, and simplicity in the system.

18. Bridges do not have types.

The final feature of Bridges is that they do not have types. Unlike Sources and Targets, which have specific types that help to differentiate them, Bridges do not require nor have any type classification. This allows for a more flexible and adaptable system, as Bridges can be created and superseded without being constrained by any specific type or category. Instead, the focus is on the relationship between the Source and Target that the Bridge represents, rather than any predefined attributes or classifications. This provides a more efficient and streamlined approach to managing agreements and relationships within a business process, as there is no need to create and maintain a separate set of types or categories for Bridges.

This lack of types also allows for greater scalability and extensibility, as the system can easily accommodate new and evolving types of agreements and relationships without needing to modify the underlying structure or framework. Additionally, it enables the system to be more easily customized to the specific needs and requirements of individual businesses, as there are no rigid preconceptions or limitations on what types of agreements or relationships can be managed through Bridges.

So, Bridges do not have types, which allows for a more flexible and adaptable system that can easily accommodate new types of agreements and relationships without needing to modify the underlying structure or framework. This lack of types also enables the system to be more easily customized to the specific needs and requirements of individual businesses.

In the complex world of QBD, the intricate network of Bridges, Sources, and Targets is not set in stone. There are specific roles that hold the power to invalidate a bridge, thus altering the landscape of connections and relationships. This power rests in the hands of four specific roles:

1. Related Owner: This is the individual or entity who is the owner of the Origin that is linked to the Target via the bridge. As the owner, they have an inherent right to question the validity of the bridge, as it directly impacts their Origin and potentially their overall stake in the business process.
2. Validator: The Validator is an independent party who is tasked with verifying the legitimacy and accuracy of the bridge. If they find inconsistencies, discrepancies, or any form of irregularity during their audit, they have the power to invalidate the bridge.
3. Authority: This is the party responsible for the Target linked by the bridge. The Authority has the mandate to ensure that all Bridges linked to their Target are valid and are serving the intended purpose. If they find that this is not the case, they can invalidate the bridge.
4. Governor: The Governor is the overarching authority that oversees the entire QBD system. They are responsible for maintaining the integrity of the system, which includes ensuring the validity of all Bridges. If they find a Bridge that does not meet the system's standards or rules, they have the ultimate power to invalidate it.

It's essential to understand that this power is not wielded arbitrarily; it serves to maintain the integrity and reliability of the QBD system, ensuring that it continues to function optimally and truthfully represent the business process it models.

Bridge Durability

The Durability of a Bridge is a property that specifies its lifespan. Established initially at the Target level as the "Bridge Durability" property, thc Durability can be set to either Timeless or Temporal. If the Target's Bridge Durability property is set to Timeless, then when creating a specific Bridge, the Broker has the choice to either maintain the Timeless setting or to define a specific Lifespan for the Bridge.

1. **Timeless**: If the Durability of a Bridge is set to Timeless, this Bridge will persist indefinitely, with its validation continuing into perpetuity unless externally invalidated by a designated authorized party.
2. **Temporal**: The Broker can also choose to set a specific Lifespan for the Bridge. In this case, the Durability of the Bridge is set to Temporal, and the lifespan is set using a duration (count) and an increment (like seconds, days, eons). For example, a Broker could establish a Bridge with a lifespan of 5 years. Once this time period elapses, the Bridge automatically becomes invalid. This can be useful in scenarios where the agreement between parties is meant to last for a specific duration.

In music, musicians often reference time signatures like three-four time (as in a waltz), or four-four time (common in hip-hop). But have you ever considered what these time signatures mean? A time signature in music is essentially a fraction: the numerator represents the number of beats in a measure, and the denominator indicates the note value that represents one beat. For example, in three-four time (3/4), there are three beats per measure and the quarter note gets the beat. Similarly, in twelve-eight time (12/8), there are twelve beats per measure, but the eighth note gets the beat.

The concept of Durability in QBD draws a parallel to this musical notation. Durability, a measure of a Bridge's lifespan, consists of two parts, similar to a musical time signature. The 'Duration' functions as the numerator, representing the count of measures in this lifespan 'signature', while the 'Increment' serves as the denominator, signifying the unit that gets the beat - for instance, years, days, or milliseconds. So, in QBD, Durability could be represented as '5 years', similar to how we might express '5/4' in music, indicating a lifespan of five years for the Bridge.

It's important to note that if the Bridge Durability of a Target is set to Temporal, a Broker cannot override this to set the Bridge to Timeless. The Broker must comply with the lifespan determined at the Target level.

The Durability property of a Bridge gives Brokers the flexibility to tailor Bridges according to the needs of various scenarios. Whether the Bridge needs to persist indefinitely or be valid for a specific duration, Quantum Business Dynamics™ has it covered. The robust design of the system allows for an incredibly versatile and adaptable structure, catering to a wide range of use cases and applications.

If a specific Bridge's Durability Lifespan lapses for a given agreement, the Bridge's state will be changed from valid to invalid. A message of this state change will be published and any subscribers to that Particle will be notified and can take action as desired.

For example, if a pilot's license to fly commercially is based on a combination of Bi-links, such as current hours, number of touch-and-go's, insurance and annual physical, among other requirements, if one of these were to laps, perhaps their security badge to swipe to enter a twin engine might be declined.

Bridge Venue

Just as the Durability of a Bridge defines its lifespan, the concept of Bridge Venue in Quantum Binary Dynamics™ (QBD™) adds an essential layer of control and clarity to the resolution of potential disputes related to Bridges brokered to specific Targets.

As a Broker, you may be familiar with the term "Venue" from the legal context, where it refers to the specific geographical location or jurisdiction where legal actions or disputes should take place. In the context of QBD, Bridge Venue Selection allows the Authority, who sets up the Target, to determine the jurisdictions where legal actions related to Bridges should be conducted.

When a Target's Bridge Venue property is set to "Venueless," which is the default setting, Bridges can operate without any predefined jurisdictional constraints. In such cases, legal actions or disputes arising from the Bridges will be subject to the existing laws and regulations of the parties involved. This may be the case for many Targets, and in such instances, you won't see any controls related to Bridge Venue during the Bridge creation process.

However, if the Authority has chosen to set one or multiple Bridge Venues for a specific Target, you, as a Broker, will be presented with a list of the available venues during the Bridge creation process. Here's where your choice comes into play. You can select one or more venues from the list provided by QBD, adhering to the constraints set by the Authority.

It's essential to understand that if the Authority has not specified any Bridge Venues, you do not have the option to add them. The lack of Bridge Venue selection indicates that the Authority has decided to leave the jurisdictional considerations open, allowing Bridges to operate without any predefined venue constraints.

On the other hand, if you find a Target that aligns with your preferences, you can take advantage of the predefined Bridge Venues set by the Authority. By selecting a pure subset of these venues for your Bridge, you ensure alignment and compliance with the Target's jurisdictional preferences.

In situations where the set of Bridge Venues offered by the Authority doesn't align with your needs or preferences, you have the freedom to either look for another Target with different venue options or create a new Target yourself, tailoring it to your specific requirements.

In summary, Bridge Venue Selection provides you with the flexibility to navigate jurisdictional considerations and make informed choices when brokering Bridges. It empowers you to align with the Authority's vision for the Target while also providing the autonomy to find the right fit for your business objectives and legal comfort.

So, as you explore the realm of Bridge Venue in QBD, remember that your choices play a pivotal role in determining the jurisdictions where legal actions related to Bridges will take place. Embrace this aspect of control and transparency, and let it guide you in brokering Bridges that shape a thriving and interconnected business ecosystem within Quantum Binary Dynamics™.

Bridges in Summary

In conclusion, Bridges provide a powerful tool for businesses to establish and maintain accurate, auditable, and verifiable relationships between Origins and Targets. They offer benefits such as immutable records, easy traceability, and the ability to update agreements as circumstances change.

Industries such as wholesale trade and public administration can also benefit from the use of Bridges. For example, in the wholesale trade industry, Bridges can be used to establish and maintain relationships between suppliers and buyers, ensuring accurate record-keeping and facilitating easier verification of agreements. Similarly, in the public administration industry, Bridges can be used to establish and maintain relationships between government agencies and citizens, providing a more transparent and accountable government.

As noted by Thomas Redman, author of "Data Driven: Creating a Data Culture," "When we can't tell where data came from or how it was created, we can't trust it." Bridges provide a solution to this problem by ensuring accurate record-keeping and easy verification of agreements, which in turn builds trust in the data and the business processes that rely on it.

With Bridges, businesses can create a system that is transparent, auditable, and trustworthy, which in turn can help them to reduce risk, increase efficiency, and improve the overall quality of their business processes.

As we move into the next chapter on Binary Link tokens, we will explore how Bridges play a crucial role in the creation and management of these tokens, and how they can be used to establish and maintain accurate, auditable, and verifiable relationships between Origins and Targets. So, let's dive into the world of Binary Link tokens and explore the many possibilities they offer for businesses looking to take advantage of the benefits of QBD.

Binary Links

In the ever-evolving landscape of business, there is one thing that remains constant: trust, transparency, and security are paramount. This is especially true for industries such as mining and utilities, where agreements and transactions are complex and involve multiple parties. That's where Binary Links come in - a revolutionary tool in the world of Quantum Business Dynamics™ that provides a secure and reliable way to establish and maintain trust between parties.

With Binary Links, executives in these industries can streamline their transactions and agreements, from the exploration and mining of minerals to the generation and distribution of energy. Imagine a mining company using Binary Links to bridge agreements with the government and local communities for sustainable mining practices. Or a utility company using Binary Links to establish and maintain agreements with regulators, suppliers, and customers for a seamless and efficient energy grid.

Through the following 19 features, we will dive into the world of Binary Links and explore their unique attributes, from their immutable and permanent record to their ability to bridge Origins and Targets. By the end of this guide, you'll have a comprehensive understanding of how Binary Links can benefit your business and enhance the trust, transparency, and security of your transactions and agreements. So, fasten your seatbelts, and get ready to discover the treasure trove of Quantum Business Dynamics™ with Binary Links.

1. Binary Links are always a point of Origin.

The first feature of Binary Links is that they serve as Origins for downstream agreements, despite being created as a result of a previous bridging of an Origin to a Target. In QBD methodology, an Origin is a particle that can be bridged to one-to-many Targets, which for each creates a new relationship or represents the start of a new agreement between two parties. Once a Binary Link is created, it now acts as an Origin itself, which, like Sources, can then be used as needed to initiate new agreements between the parties involved.

Although it represents the initiation of a previous agreement, it can also serve as a starting point for future agreements with other Targets. While its creation does not add new rules governing the initial agreement, once established as an Origin, it can be instrumental in generating new sets of rules governing downstream agreements.

Binary Links are unique in their ability to represent two parties that are entering into a new agreement, and as a result, they are crucial in ensuring that the agreement is accurately represented and understood. This feature ensures that the agreement is clearly defined and that both parties understand the terms and conditions of the agreement.

In practical terms, Binary Links function as a way to establish trust between two parties entering into a new agreement. They provide a record of the terms of the agreement, including the obligations of each party and the consequences of failing to meet those obligations.

For businesses, the use of Binary Links can be particularly valuable in establishing trust and transparency in their relationships with partners, suppliers, and customers. By using Binary Links to document their agreements, businesses can ensure that both parties have a clear understanding of the terms of the agreement and can rely on the immutability of the record to verify the terms at any time in the future.

Overall, the first feature of Binary Links highlights their critical role in establishing new agreements and building trust between parties. By creating a clear record of the terms of the agreement, Binary Links can help businesses establish transparency and accountability in their relationships, ultimately leading to stronger partnerships and more successful outcomes.

2. Binary Links are the result of some Bi-link chaining event where a Bridge was created.

Binary Links are the result of a Bi-link chaining event where a Bridge was created. This Binary Link particle can serve as a new Origin for any future agreements involving two parties, allowing for the creation of a chain of Bi-links. These chains can be used to create complex networks of relationships, providing a traceable and auditable history of all agreements and transactions within the network.

This can be particularly valuable in industries that require a high level of trust and transparency, such as finance, healthcare, and supply chain management. With QBD methodology, businesses can ensure that their agreements are secure, immutable, and easily verifiable, reducing the risk of fraud and disputes. As one source noted, "Blockchain can help the industry create a more reliable, transparent, and secure transactional ecosystem. This creates a level of trust between parties that might not have existed before, and that could facilitate growth and the opening of new markets."

For businesses, the ability to track and audit agreements in a transparent and immutable way can be crucial for maintaining trust with customers and partners. The use of Binary Links as the result of Bi-link chaining events through Bridges can provide a robust system for managing complex business relationships and ensuring compliance with regulations. In industries such as finance, healthcare, and real estate, where trust and compliance are critical, the use of QBD methodology and Binary Links can provide a competitive advantage.

Furthermore, the creation of Bi-link chains can also streamline and automate the creation of new agreements. Once a Binary Link has been established, it can be used as an Origin to initiate future agreements with new Targets, allowing for quick and efficient relationship-building between parties. This can save time and resources for businesses, enabling them to focus on other important tasks and initiatives.

In summary, the creation of Binary Links as a result of Bi-link chaining events through Bridges can provide businesses with a transparent and immutable system for managing complex agreements, ensuring compliance, and automating the creation of new relationships. By adopting QBD methodology and leveraging the power of Binary Links, businesses can stay ahead of the curve and build trust with their customers and partners.

3. All prior validations encapsulated in the history of a Binary Link's relationships can be subsequently verified over and over again.

One of the key benefits of QBD and Binary Links is the ability to verify and authenticate the validity of an agreement through the entire history of its relationships. Because the validation of the Source is done at the time of its creation, any subsequent Bi-Links can be trusted for that validated as well. This means that parties can have confidence that the agreement is reliable, immutable, and transparent throughout its entire history.

This feature of Binary Links provides an additional layer of security and trust for parties involved in agreements. It also enables third parties with proper authorization to verify the validity of the agreement at any time, which can be especially useful in legal or regulatory situations. As such, Binary Links can serve as a powerful tool in promoting transparency and accountability in various industries.

For example, in healthcare, the use of Binary Links can ensure that patient data is securely shared between parties with proper authorization, while also enabling verification of the validity of agreements between healthcare providers and patients. In finance, Binary Links can help ensure that financial transactions are securely and transparently conducted between parties, while also enabling third-party verification of the validity of agreements.

In finance, Binary Links can be used to ensure the validity and accuracy of financial transactions, including the transfer of funds and the tracking of assets. The transparency provided by QBD can help to reduce the risk of fraud, as well as provide greater confidence and trust between parties.

In the legal field, Binary Links can be used to ensure the validity and enforceability of contracts and other legal agreements. The ability to verify the prior validations of a Binary Link provides a level of assurance and security that is essential in legal proceedings. This can help to reduce the risk of litigation and other legal disputes, as well as improve the overall efficiency of the legal system.

Overall, the ability to verify and authenticate the validity of an agreement through the entire history of its relationships is a significant benefit of Binary Links. By leveraging this feature, businesses can promote transparency, security, and trust in their agreements and transactions, while also enabling third-party verification of the validity of these agreements.

4. Binary Links have five attributes that are used in their creation.

The fourth feature of Binary Links is that they have five attributes used in their creation. Four of these attributes are the same as Sources, which are the Identifier selected by the Validator, Description which is automatically set by the QBD system, Validation Type which is automatically set to "Binary Link," and Image selected by the Validator. In addition to these four attributes, Binary Link tokens have an additional attribute that identifies the Bridge responsible for its creation.

This unique attribute allows for traceability and accountability in the creation of Binary Links. It ensures that any disputes or questions regarding the validity or terms of an agreement can be traced back to the specific Bridge that created the Binary Link. As a result, the parties involved can be held accountable for their actions, and there can be no confusion about the terms and conditions of the agreement.

The use of Binary Link attributes in QBD methodology also ensures that the creation of Binary Links is standardized and consistent across the system. This standardization allows for easy integration with other systems and technologies, making it easier for businesses to adopt and implement QBD. Additionally, the standardization of attributes enables automated processes to be developed and implemented, reducing the risk of human error and increasing efficiency.

The unique attribute that identifies the Bridge responsible for creating the Binary Link also facilitates the process of superseding Bridges, as discussed in the previous feature. The ability to trace a Binary Link back to its originating Bridge enables the superseding Bridge to invalidate the previous agreement and create a new one, ensuring that the agreement remains transparent and accountable.

In summary, the five attributes of Binary Links ensure standardization, traceability, and accountability in the creation of agreements. These attributes enable easy integration with other systems, reduce the risk of human error, and facilitate the process of superseding Bridges. By adopting QBD and Binary Links, businesses can benefit from increased transparency, security, and trust in their agreements, ultimately leading to more successful and efficient business relationships.

5. The validation can be invalidated, which will break all subsequent Bridges to other Targets from that Binary Link.

Binary Links are unique in that their validation can be invalidated, which can cause all subsequent Bridges to other Targets from that Binary Link to break. This is because the Binary Link is dependent on its validation to maintain its integrity and trustworthiness as an Origin particle. If its validation is no longer considered valid, any agreements that were made based on that Binary Link will also lose their validity.

While this may seem like a potential drawback, it actually serves an important purpose in maintaining the overall trustworthiness of the QBD methodology. By allowing for the possibility of invalidation, QBD ensures that any agreements made using Binary Links are only valid as long as the conditions under which they were agreed upon remain true. This prevents parties from continuing to act based on outdated or inaccurate information, which could lead to misunderstandings or even legal disputes.

However, it's important to note that the invalidation of a Binary Link's validation should not be taken lightly. This action should only be taken if there is evidence that the initial validation was made in error or based on incorrect information. Once a Binary Link's validation is invalidated, any subsequent Bridges to other Targets from that Binary Link will no longer be considered valid, and parties will need to renegotiate their agreements using a new Origin particle.

Overall, Binary Links provide a secure and trustworthy way for parties to enter into agreements, with the added benefit of being able to maintain and verify those agreements over time. By allowing for the possibility of invalidation, QBD ensures that parties can act with confidence in the knowledge that any agreements made using Binary Links are based on accurate and up-to-date information.

6. When the Bridge that yielded a Binary Link is subsequently invalidated, it is superseded by a new bridge.

The sixth feature of Binary Links is that when the Bridge that originally yielded a Binary Link is invalidated, a new Bridge will supersede it. In the case that the original invalidation is later corrected, a completely new Binary Link will be generated from the new Bridge. The previous Binary Link token will still exist and can serve as evidence of the state of the agreement between the parties during the time period from the original agreement until its invalidation.

It's crucial to ensure the validity of all Bridges in the QBD methodology, as any invalidation of a Bridge can have ripple effects on downstream agreements. However, when a Bridge is superseded by a new one, the resulting new Binary Link token can eliminate confusion and ambiguity in the documentation of the agreement. Relying parties subscribed to the state of the underlying Bridge will receive notice of its change, and when a new Binary Link is generated, trust can be high in re-establishing downstream agreements.

For example, let's say a nurse used a Binary Link token representing the RN degree to get a discount at a nursing supply store. If her degree is later found to be invalid, the store may revoke her discount. However, when the error is corrected and a new Binary Link token representing her continued good standing is generated, the store is likely to reinstate her discount without requiring her to associate the new Binary Link with the same Target again. On the other hand, if the hospital that suspended her due to the invalidation of her degree now wants to reinstate her, they will likely require her to bridge the new Binary Link to their HR Target before allowing her back on hospital grounds.

This feature of Binary Link particles emphasizes the importance of proper record-keeping and documentation built into QBD to ensure that any changes to the state of an agreement are accurately recorded and can be traced if necessary. With clear and concise records of all Bridges and Binary Links, including the reasons for any invalidations or superseding Bridges, businesses can maintain the integrity of their agreements and minimize confusion and ambiguity in their relationships.

Furthermore, this feature showcases the flexibility of the QBD methodology. If a Bridge is found to be invalid, a new Bridge can be created to replace it, allowing the parties to continue their agreement without interruption. This adaptability is a valuable asset for businesses that may encounter unexpected obstacles or changes in their agreements.

7. Binary Links act additionally as Origins after creation and as such naturally seek Targets to the benefit not only of the business, but also businesses in entirely different industries.

Binary Links, as we have discussed, are the yield of bridging an Origin to a Target and serve as a new Origin themselves. Once created, they can be used to initiate new agreements between parties, even those in different industries. This is a powerful feature of QBD as it allows businesses to easily and securely collaborate with other businesses and industries.

For example, a healthcare provider might have a Binary Link representing their medical license. With this Binary Link, they could easily initiate a new agreement with a pharmacy to share patient information or collaborate with a medical device manufacturer to test and develop new products. Similarly, a manufacturer might have a Binary Link representing their product design patent, which they could use to collaborate with a logistics company to efficiently transport their products.

This feature of Binary Links promotes cross-industry collaboration and innovation, which can benefit all parties involved. By leveraging the secure and immutable nature of QBD and Binary Links, businesses can confidently enter into agreements with partners they might not have considered before. This can lead to new business opportunities, increased efficiencies, and ultimately, increased revenue.

It's important to note, however, that this feature also requires careful consideration of the trustworthiness of the parties involved. While Binary Links provide a secure and transparent method of collaboration, it is still important to thoroughly vet potential partners and ensure that all agreements are properly documented and recorded. By doing so, businesses can fully take advantage of the benefits of QBD and Binary Links while mitigating potential risks.

8. Binary Links relationships to downstream Targets can be verified repeatedly.

One of the key benefits of QBD and the use of Binary Links is the ability to verify the relationships between parties and agreements over time. Once a Binary Link is created, it acts as a verified Origin that can be used to establish new agreements with downstream Targets. This means that any business that enters into an agreement with another party can be confident that the agreement is valid and that the downstream Targets can be trusted.

This feature is especially important in industries where trust is paramount, such as finance, healthcare, and legal services. By using Binary Links to establish and verify agreements, businesses can reduce the risk of fraud and ensure that all parties involved in the agreement are acting in good faith. Additionally, this can lead to more efficient and effective business practices, as parties can focus on executing the terms of the agreement rather than spending time verifying the validity of the relationship.

The ability to verify relationships over time also has benefits beyond the initial agreement. As businesses grow and expand, they may need to establish new agreements with different parties. By using Binary Links as Origins, they can quickly and easily establish trust and confidence with new parties, knowing that the relationship can be verified over time. This can help businesses to expand more quickly and effectively, as they can establish new partnerships and agreements with confidence.

It is also worth noting that the ability to verify relationships over time can be particularly useful in industries that are subject to regulatory oversight. By using Binary Links to establish and verify agreements, businesses can provide regulators with clear and concise documentation of their relationships and agreements. This can help to streamline the regulatory process and reduce the risk of penalties or fines for non-compliance.

9. Binary Links can only point at Targets, either through the yield back to the Target instrumental in its creation or forward through future Bridges to additional Targets.

Binary Links are unique in their ability to link a specific Origin to a specific Target. Once created, a Binary Link can only point from Targets either through the yield back from the Target instrumental in its creation or through future Bridges to additional Targets. This one-way relationship ensures the integrity of the agreement represented by the Binary Link, as it can only be used for the intended purpose and cannot be repurposed for any other agreement.

This feature of Binary Links makes them ideal for industries where security and privacy are of utmost importance, such as healthcare, finance, and legal services. In these industries, ensuring that agreements are only used for their intended purposes and that sensitive information is only shared with authorized parties is crucial. The unidirectional nature of Binary Links helps to ensure that these requirements are met, as the links can only be used to connect the Origin to the intended Target or additional authorized Targets.

Additionally, the unidirectional nature of Binary Links can be useful for businesses looking to establish strong partnerships and collaborations with other companies. By creating a Binary Link between two companies, the link can be used to establish trust and facilitate secure data sharing, while ensuring that the data is only used for the intended purposes. This can help businesses to expand their reach and build mutually beneficial relationships with other companies in their industry or even in entirely different industries.

In summary, the unidirectional nature of Binary Links ensures that the agreements they represent can only be used for their intended purposes and cannot be repurposed for any other use. This makes them a valuable tool for industries where security and privacy are paramount, as well as for businesses looking to establish strong partnerships and collaborations with other companies.

10. The only way for a Binary Link to be related to a future Target is by a Bridge.

Binary Links are truly unique entities in the Quantum Business Dynamics™ framework. One of their most intriguing features is that they can only relate to future Targets through Bridges. This simple rule may seem like a limitation, but in fact, it provides incredible flexibility to the QBD fabric.

Consider the example of an oil well project. There may be multiple Targets associated with the project, such as one for the project itself and another for the working well. Different Bi-link chains may be associated with these Targets, such as a title chain for the families that have a stake in the project or a certification chain for the contractors involved in the work.

Thanks to the rule that Binary Links can only relate to Targets through Bridges, these different chains can intersect at the same Targets and yet result in different outcomes. For example, the family's share Bi-link may bridge directly to the Target of the working well, resulting in a new token for shares - the form of payment they will receive for the use of their title. On the other hand, a contractor may bridge to a Target of hiring a subcontractor, resulting in a token that the subcontractor can use to bridge to the Target of the working well and receive payment for their work.

By always connecting to Targets through Bridges, the QBD fabric allows for remarkable flexibility and problem resolution. And as we'll see later on, this type of intersectionality is critical for maintaining the integrity of the QBD system.

The fact that Binary Links can only relate to Targets through Bridges is a fundamental rule of QBD that makes this elegance in the fabric possible. So, while it may seem like a limitation, it actually opens up a world of possibilities for businesses that adopt QBD.

11. Binary Links must have a unique user-friendly handle.

In a world where technology is constantly evolving, the need for simplicity and ease of use has become increasingly important. Binary Links address this need by requiring a user-friendly handle. This handle, which consists of a 3-character-4-character pairing, allows users to quickly and easily identify a specific Binary Link token. The handle is similar to old phone numbers, where each character can be alphanumeric.

The user-friendly handle of Binary Links is an important feature that helps to simplify the complex nature of Quantum Business Dynamics™. It allows users to easily identify and track Binary Link tokens, even in a large and complex system. This feature is particularly useful for businesses that deal with a large volume of agreements and contracts.

Furthermore, the alphanumeric nature of the handle provides an additional layer of security. By using a combination of letters and numbers, the handle is much more difficult to guess or hack. This helps to ensure the integrity of the agreement represented by the Binary Link, as well as the overall security of the QBD system.

Overall, the user-friendly handle of Binary Links is a key feature that helps to make the QBD system more accessible and secure. It is a simple yet effective way to streamline the process of identifying and tracking Binary Link tokens and provides an added layer of security to the system.

12. When created, Binary Links naturally generate new attributes.

When a Binary Link is created, it generates several new attributes that are used to maintain its integrity and track its history. These attributes include the date and time of creation, the Validator responsible for its creation, and the Bridge used to create the link.

The date and time of creation is a crucial attribute as it allows for the tracking of the history of the Binary Link. This information can be used to verify the validity of the agreement represented by the Binary Link at any given time. The Validator responsible for the creation of the Binary Link is also an important attribute as it provides accountability and ensures that the agreement is being created by a trusted party.

The Bridge used to create the Binary Link is another important attribute that is generated when the link is created. This attribute is used to maintain the relationship between the Origin and the Target, and to ensure that any future agreements related to the original agreement are properly recorded and tracked.

Additionally, Binary Links can also generate other attributes based on the specific needs of the business or industry using them. These can include information about the type of agreement being represented, the duration of the agreement, and any additional terms or conditions.

Overall, the generation of these attributes is an essential aspect of Binary Links and QBD. It allows for the tracking of agreements over time, ensures accountability and trust, and allows for the customization of the system to fit the needs of any particular industry or situation.

13. Like Sources, Targets, and Bridges, there's no way to suss out a Source's relationships or even that it is a Source from publicly available data because all particles share the exact same public attribute footprint.

One of the unique features of the QBD methodology is that the public attribute footprint of a Binary Link is identical to that of a Source, Target, or Bridge. This means that there is no way to identify a Binary Link or its relationships from publicly available data. This is an intentional design choice in order to protect the privacy and security of the parties involved in a QBD agreement.

As explained in earlier features, Binary Links act as an Origin and relate to Targets through Bridges. These relationships are only visible to the parties involved and can be verified through the use of the private key associated with the Origin. This provides an additional layer of security and trust in the QBD methodology.

However, it is important to note that while the privacy and security of QBD are strengths of the methodology, it also requires a level of trust between parties. As with any agreement, it is crucial to thoroughly evaluate the trustworthiness of the parties involved before engaging in a QBD agreement. Additionally, proper documentation and record-keeping are important to ensure that any changes to the state of the agreement are accurately recorded and can be traced if necessary.

14. Further like the other particles, the record of their attributes is immutable and permanent, good for reproof, verification, and even legal nonrepudiation.

Binary Links are not just a useful tool for businesses looking to improve their transparency and security - they are also an invaluable resource when it comes to maintaining an accurate record of agreements and transactions. One of the key features of Binary Links is their immutable and permanent nature. Just like Sources, Targets, and Bridges, the record of Binary Links' attributes cannot be altered once they are created. This is due to the blockchain technology used by QBD, which ensures that all transactions and agreements are permanently recorded and cannot be tampered with.

This feature of Binary Links has several important implications for businesses. First and foremost, it means that Binary Links can serve as a reliable source of proof and verification in the event of any disputes or legal proceedings. Because the record of their attributes is immutable, it is impossible for anyone to alter or dispute the terms of an agreement represented by a Binary Link. This provides businesses with a powerful tool for resolving disputes and proving the validity of their agreements.

In addition to their use as a source of proof and verification, Binary Links also offer businesses the benefit of legal nonrepudiation. Nonrepudiation refers to the ability to prove that a certain party was involved in a transaction or agreement and cannot deny their involvement. By recording all attributes of Binary Links in an immutable and permanent record, businesses can establish a clear and indisputable record of their agreements and transactions. This can be especially valuable in industries where disputes and legal challenges are common.

Overall, the immutability and permanence of Binary Links' attributes make them an essential tool for businesses looking to improve their transparency, security, and reliability. By providing a clear and indisputable record of their agreements and transactions, Binary Links offer businesses a powerful tool for resolving disputes, establishing legal nonrepudiation, and maintaining trust with their partners and customers.

15. Verifications of any of the attributes also become immutable and permanent, good for reproof, verification, and even legal nonrepudiation.

Binary Links offer the same level of immutability and permanence for verifications of their attributes as for the record of their creation. This means that once a verification is made, it becomes a permanent part of the Binary Link's history and cannot be altered. This level of permanence and immutability is essential for businesses operating in industries where transparency and accuracy are critical, such as finance, healthcare, and legal services. The ability to provide a permanent and immutable record of verifications can serve as evidence in legal disputes and help businesses protect themselves from fraudulent claims. It also provides a level of trust and assurance for all parties involved in a transaction or agreement, knowing that the verification process and its results cannot be tampered with or altered.

16. Binary Links are singular unto themselves and there cannot be a copy of them.

Binary Links are unique in their nature, in that they are singular entities that cannot be copied or duplicated. Each Binary Link is created with a specific set of attributes, and these attributes are immutable and permanent. This means that any changes made to the Binary Link will be recorded and can be traced back to the original agreement.

The fact that Binary Links cannot be copied is a key feature of the QBD methodology. It ensures that the integrity of the agreement represented by the Binary Link is maintained, and that there is no risk of the agreement being misrepresented or misunderstood. This also means that there can be no confusion or ambiguity regarding the terms of the agreement.

In addition, the uniqueness of Binary Links provides an added layer of security and trust. Since each Binary Link is singular unto itself, it cannot be duplicated or falsified. This means that businesses can be confident that the Binary Links they rely on are authentic and represent valid agreements.

Overall, the singularity of Binary Links is an essential feature of the QBD methodology. It ensures that the integrity and authenticity of agreements are maintained and provides a high level of security and trust for businesses.

17. Binary Links are uniquely possessable, as they are represented as digital tokens that can be withdrawn to any standard wallet that supports an underlying network.

Binary Links are unique in that they are not only singular unto themselves, but they are also uniquely possessable. Each Binary Link is represented as a digital token that can be withdrawn to any standard wallet that supports the underlying network. This means that the token can be transferred between parties just like any other digital asset, allowing for ownership and control to be easily transferred. The use of digital tokens for Binary Links also makes them highly portable and easy to store, without the need for physical documents or paperwork. This feature is especially beneficial for businesses that frequently engage in transactions with multiple parties, as it allows for a streamlined and efficient transfer of ownership and control. Additionally, the use of digital tokens for Binary Links ensures that they are tamper-proof and cannot be duplicated, providing an extra layer of security for all parties involved in the agreement. With this level of trust and security, businesses can have peace of mind that their agreements are being upheld and enforced, and that they have a reliable and immutable record of all their transactions.

18. Binary Links may only have one type assigned to them, and it is dictated by the Probabilistic/Deterministic settings of its Target.

This feature ensures that there is no confusion or ambiguity about the nature of the particle, and it also enables streamlined verification processes. Further, it ensures that they are Origins that can be bridged to other appropriate Targets, and that they are instrumental in establishing and maintaining trust in business relationships. By assigning the Binary Link type to these particles, it becomes clear to all parties involved that this is a unique and immutable record of an agreement between an Origin and a Target.

This rule is essential for the smooth functioning of the QBD methodology. By limiting Binary Links to one type, it becomes easier to verify and authenticate them, ensuring that they are being used in a trustworthy manner. The transparency and immutability of Binary Links make them an ideal tool for establishing trust in business relationships, and the fact that they can be easily transferred to any standard wallet that supports the underlying network makes them a versatile and convenient option for businesses of all types.

One of the benefits of this feature is that it ensures that businesses can maintain a clear and concise record of their agreements, and it also enables them to easily verify the authenticity of those agreements if necessary. This is particularly important in industries where transparency and trust are paramount, such as the financial and healthcare industries. Binary Links provide a high level of transparency and accountability, which can help to prevent fraud and other types of malfeasance.

Overall, the fact that Binary Links can only have one type assigned to them is a crucial feature of the QBD methodology. It ensures that these particles are used in a trustworthy and transparent manner, and it also enables businesses to maintain a clear and concise record of their agreements. Whether you are a small business owner or the CEO of a large corporation, the immutability and transparency of Binary Links can help you to establish and maintain trust in all of your business relationships.

19. A Binary Link Identifier must be unique.

Binary Links are identified by a unique identifier, which must be distinct from all other Binary Links. This is crucial in ensuring the integrity and accuracy of the QBD system. Without unique identifiers, it would be difficult to distinguish between different Binary Links and their respective agreements, leading to potential errors or confusion in the system. Therefore, it is important that a Binary Link Identifier must be unique, and this rule is strictly enforced in the QBD methodology.

Binary Links in Summary

As we wrap up our discussion on QBD Particles that make up the Bi-link Fabric, we hope you have gained a deeper understanding of the importance of Sources, Targets, Bridges, and Binary Links in building and maintaining trust, transparency, and security in today's digital landscape.

By understanding the critical role that these particles play in the QBD methodology, executives in various industries can take advantage of this powerful tool to solve typical problems they might encounter in their businesses.

For example, in the mining and utilities industry, where project management and safety are paramount concerns, implementing QBD can ensure that all parties involved in a project have access to the same, accurate information, reducing the risk of errors, accidents, and delays.

Similarly, in the healthcare industry, where compliance and privacy are essential, QBD can provide a reliable method of verifying credentials and ensuring that sensitive information is shared only with authorized parties.

Now that you have learned about Sources, Targets, Bridges, and Binary Links, the next step is to understand how to use this methodology with your team. In the next section, we will discuss how to implement QBD in your organization, including how to train your team, how to choose the right technology, and how to create a culture of trust and transparency.

Quantum Business Dynamics™ is a powerful tool that can transform the way businesses operate in the digital age. By understanding and implementing the principles of Sources, Targets, Bridges, and Binary Links, executives in any industry can improve transparency, trust, and security, and achieve greater success in their business endeavors.

So, let's move forward together, and explore how to take advantage of the full potential of Quantum Business Dynamics™. In the next section, we will delve deeper into the practical aspects of implementing QBD in your organization and provide you with the tools you need to succeed.

Particle Comparison/Contrast

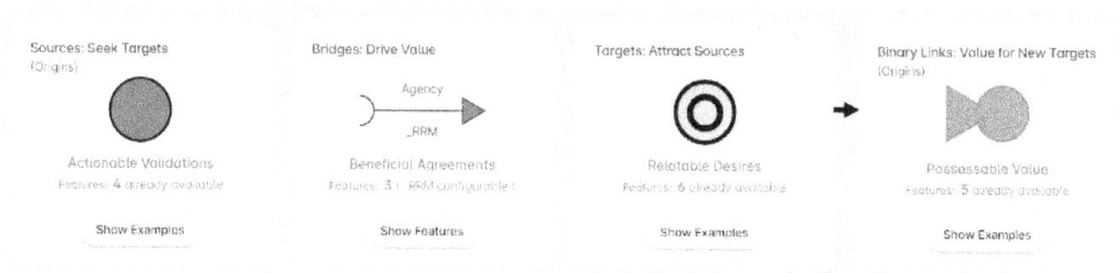

Quantum Business Dynamics™ (QBD™) methodology introduces the concept of particles as a fundamental building block in its architecture. In this section, we will provide a comprehensive comparison and contrast of particles, highlighting their commonalities and differentiators. By understanding these aspects, one can better appreciate the nuances of particles and how they contribute to the overall QBD framework.

Particle Commonalities:

At their core, particles share several common features that make them a crucial component of the QBD system. These features are essential in providing a consistent and secure way of managing and processing data across various industries and use cases.

1. Unique Identifiers: Each particle in QBD carries a unique identifier that distinguishes it from other particles within the system. This allows for efficient tracking, retrieval, and management of data throughout the entire QBD framework.
2. Cryptographic Security: All particles employ advanced cryptographic techniques to ensure the security and integrity of the data they hold. This includes the use of digital signatures and hashing algorithms, which help protect against unauthorized access and tampering.
3. Bi-link Relationships: Particles are interconnected through Bi-link relationships, which serve as the foundation for creating complex data structures in QBD. These relationships enable the establishment of trust and verifiability across the entire data chain, allowing for secure and transparent data sharing among multiple entities.
4. Consensus Mechanisms: To maintain data integrity and consistency, particles rely on consensus mechanisms that validate and confirm the authenticity of their Bi-link relationships. This helps create a robust and tamper-resistant data structure, providing a high degree of trust in the system.

Immutable Public Features:

- Class ID (either Series or Topic Id)

- Serial Number
- CID (Content Identifier)
- Image (optional)
- Alt Image (optional)
- History (quantum resistant hash)
- Veracity (quantum resistant hash)

Immutable Confidential Features:

- Identifier
- Description
- Front End Identifier
- Messages
- State

All particles:

- Have a publicly available Series number, Serial number, and CID hash
- Have a confidential Handle that can be shared with proper authorization
- Have hashed features for Veracity, History, and CID
- Have patternless pubic attributes, making them indistinguishable from one another
- Have attributes used to create patternless particles, increasing the security and transparency of the QBD system

Particle Differentiators:

While particles share several commonalities, they also exhibit unique features that differentiate them based on their specific functions and use cases within the QBD system. These differentiators play a vital role in customizing and adapting the QBD framework to cater to the diverse needs of various industries.

1. Data Types: Particles can hold different types of data, depending on their purpose within the QBD system. For instance, some particles may store personal information, while others might hold transactional data or even metadata about other particles. The ability to store different data types makes particles versatile and adaptable to various applications.
2. Access Controls and Permissions: Particles can have varying levels of access controls and permissions, which are determined by their specific use cases and the sensitivity of the data they hold. This allows for granular control over data access, ensuring that sensitive information is protected and only shared with authorized parties.

3. Customization and Flexibility: Particles can be customized to fit the unique requirements of individual businesses or industries. This includes the ability to create custom workflows, business rules, and validation processes that cater to specific needs. This flexibility enables the QBD system to accommodate diverse use cases and adapt to various business environments.
4. Integration with External Systems: Particles can be designed to integrate with existing systems and technologies, facilitating seamless data exchange and interoperability. This is achieved through the use of APIs, standard data formats, and modular architecture, allowing businesses to incorporate QBD into their existing workflows without major disruptions.

Particles play a critical role in the QBD methodology, providing a consistent and secure foundation for managing and processing data across various applications. By understanding the commonalities and differentiators of particles, one can appreciate their versatility and adaptability, making them a valuable asset in implementing QBD solutions for a wide range of industries and use cases.

- Sources have a Source ID, Validation Type, and Source Hash attribute.
- Targets set the Agency, Allowable Validation Types, Vectorization, Payment for Verification, Owner of Binary Links, and Data Partners. Targets can set the Durability of all resulting Bridges to either Timeless or Temporal. Targets also have a Target ID, Target Type, and Target Hash attribute.
- Bridges do not have confidential Identifiers, Descriptions, or a Media hash. Bridges can set their own Durability when a Target has it set to Timeless. Bridges have confidential Agency, _RRM, and Broker. Bridges also have a Bridge ID, Bridge Type, and Bridge Hash attribute.
- Binary Links have a Binary Link ID, Binary Link Type, and Binary Link Hash attribute.

Important Roles

In Quantum Business Dynamics™ (QBD™), various roles are responsible for ensuring the integrity, security, and proper handling of data and information. These roles range from agents and brokers to owners and validators, each with their own specific responsibilities and duties within the QBD system. In addition to these traditional roles, QBD also includes the role of the _RRM, a nonprofit authorized agent that acts as an information rights agent for the owner of data generated from their actions on the applicable system. The _RRM works to help the owner exercise their consent elections and ensure that their data rights are protected. QBD also includes guilds, self-regulating organizations that certify _RRM candidates and provide oversight, set standards, and ensure that those working within their industry are following best practices. Together, these roles and organizations form a crucial part of the QBD ecosystem, ensuring that data is handled securely, accurately, and in accordance with established best practices.

- Agent: A person or entity authorized to act on behalf of another person or entity. In QBD, agents manage Bridges for the Owner.
- Authority: An organization or entity with the power to make decisions and enforce rules. In QBD, authorities create the Target and set its attributes.
- Broker: A person or entity that facilitates transactions between two or more parties. In QBD, brokers build Bridges as a result of an agreement between two people.
- Data Partner: A company that collaborates with another company to share data or information.
- Governor: The person who manages the whole QBD platform and owns and manages the main treasury. There is one Governor per platform. The Governor is responsible for appropriately managing the invoicing of charges for underlying services, such as blockchain gas prices, to various parties based on the rules of QBD as set in Particles by the people of appropriate roles such as Authorities when they create Targets.
- Guild: A self-regulating organization that certifies Agents and _RRMs. In QBD, the guild that manages the _RRMs is GCA3 (Global Consent Authorized Agency Association). Each industry has their own approach to what we call Guilds. Most of the time, industries are self-regulated, but in some cases, they are regulated.
- Holder: The person who currently possesses a token (Source or Binary Link) in their wallet.
- Owner: A person or entity that has legal rights and control over a particular asset or property. In QBD, the owner is the person who owns the information rights for the Source and who can provide direction to the _RRM for consent management regarding their related information rights.
- RRM: Rights Resource Manager. In QBD, an _RRM is a nonprofit authorized agent represented in the bridge, responsible for acting as an information rights agent for the owner of the data generated from their actions on the applicable system. They work to help the owner exercise their consent elections and ensure that their data rights are protected. The _RRM manages the information rights particular to

the industry represented by a bridge's agency for the Owner to ensure their consent elections are honored.

- Validator: A person or entity responsible for verifying the authenticity and accuracy of data or information. In QBD, validators create Sources as a result of validating facts. For Sources, they validate claims regarding some distinctive of identity, and for Bridges, they validate facts about an agreement between two people.
- Verifier: A person or entity responsible for checking and confirming the validity of data or information. In QBD, verifiers use existing facts discovered through handles to satisfy a question or request regarding the status of relationships. There can be many verifiers, both human and machine.

Y'all ain't gonna believe this $#!T

Now that you have a grounded understanding of the Particle basics, it's time to elevate a bit for context and think through practical applications and the vast benefits that are waiting for you when you adopt QBD. It's like going through a too-brief primer on parachutes, and then finding yourself at the open door of a Cessna, cold wind blasting across your exposed nose and cheeks as you take one last gulp, and steeling your nerves, you jump into the endless, blue sky.

But once you take that leap, there's no going back. You will never look at your business the same way again.

The funny thing is…it's always looked like this. You have just never *seen* it before.

Now you do.

What we have basically revealed to you is that there are small, constant mechanics to the way businesses work and they behave in predictable ways, even across different disciplines and industries. Targets exist, but they operate more to your benefit if you identify them and add controls rather than letting them operate all willy-nilly. Origins are going to be attracted to Targets, but they can work more to your favor if you understand that fact and manipulate the gravitational pull by making your Targets more attractive. Agreements are going to happen throughout your business, but their resulting benefits and values will just fall on the floor if you do not capture them in ways upon which increasing value can be consistently built, relying on these predictable mechanics.

In this section, we are going to dip briefly into that sky of opportunity by examining QBD's unavoidable impacts on Generally Accepted Account Principles (GAAP), the resurrection of Philip B. Crosby's theories on Quality, and the free biproducts awaiting when our data is ruthlessly curated in an effortless way just by adopting QBD to our existing processes and systems.

Fall Into the GAAP

1. Introduction

GAAP, or Generally Accepted Accounting Principles, is a set of standardized accounting rules and guidelines that companies in the United States must follow when preparing their financial statements. GAAP aims to promote consistency, transparency, and comparability in financial reporting, making it easier for investors, regulators, and other stakeholders to understand and compare financial information across different companies.

Developed and maintained by the Financial Accounting Standards Board (FASB), GAAP covers various aspects of financial reporting, including revenue recognition, expense classification, asset and liability measurement, and financial statement presentation. The principles are based on key underlying concepts, such as accrual accounting, consistency, materiality, and conservatism.

By adhering to GAAP, companies can ensure that their financial statements provide a true and fair view of their financial position and performance. This, in turn, helps maintain trust and confidence in the financial markets and enables informed decision-making by investors and other stakeholders.

GAAP has brought numerous benefits to the business community by promoting consistency and transparency in financial reporting. However, it has also introduced certain challenges that businesses must contend with. Here are three significant challenges that GAAP has posed for the business community:

A. Complexity and frequent updates: The GAAP guidelines can be intricate and complex, making it difficult for some businesses, particularly smaller ones, to fully understand and comply with them. Additionally, the FASB regularly updates and modifies GAAP, which requires companies to stay abreast of changes and implement them in their financial reporting. This constant evolution can be time-consuming and costly for businesses to manage.

B. Subjectivity and estimates: GAAP often requires companies to make estimates and use judgment in certain areas of financial reporting, such as **assessing the useful life of assets, calculating bad debt allowances, and valuing intangible assets like intellectual property** *(more on this later in The Resurrection of Philip B. Crosby)*. These subjective decisions can introduce variability and uncertainty in financial statements, making it more challenging for investors and stakeholders to compare financial performance across companies.

C. Compliance costs and resources: Complying with GAAP can be resource-intensive, particularly for smaller businesses with limited financial and personnel resources. Implementing the necessary accounting systems, hiring or training staff to manage GAAP compliance, and engaging external auditors to review financial statements can be expensive and time-consuming. This burden can sometimes divert resources from other critical business operations.

One significant change introduced by GAAP is the increased focus on recognizing the value of intangible assets, such as intellectual property, brand recognition, and customer relationships. This shift has had a considerable impact on business practices, as it has forced companies to reevaluate how they value, manage, and report these intangible assets. Companies now must dedicate more resources to accurately identifying, measuring, and reporting the value of their intangible assets, which can be challenging due to the inherently subjective nature of such valuations. This change has also prompted organizations to consider the strategic implications of their intangible assets and how to best leverage them to create a competitive advantage.

As you've been reading about the challenges presented by GAAP, you might be thinking about the potential role of Quantum Business Dynamics™ (QBD™) in addressing these issues. QBD, with its proactive and systematic approach to process improvement, offers a promising solution to the complexities and uncertainties faced by businesses in complying with GAAP. By promoting a culture of continuous improvement and data-driven decision-making, QBD can help organizations navigate the intricacies of financial reporting and enhance the accuracy and reliability of their financial statements. Furthermore, the emphasis on collaboration, communication, and stakeholder involvement that QBD brings can lead to a more comprehensive understanding of the value of intangible assets and a more efficient allocation of resources. The principles of QBD, when applied to financial reporting and compliance, have the potential to create a more resilient, transparent, and effective accounting environment for businesses of all sizes.

As we delve further into the relationship between QBD and GAAP, it becomes apparent that these two concepts can complement and support each other in creating a robust financial framework for businesses. In the following section, we will explore the synergies between these two approaches, examining how QBD principles can be applied to the challenges and complexities of GAAP accounting. By understanding the ways in which QBD and GAAP can work together, businesses can create a more efficient, accurate, and transparent financial reporting system that better serves the needs of all stakeholders.

2. Natural Law Applied to GAAP via QBD

Building on the powerful synergies between QBD and GAAP, we are poised to embark on a thrilling journey that transcends traditional boundaries and explores the fascinating intersection of science and accounting. As we delve into these innovative ideas and draw inspiration from the remarkable principles of chemistry and natural law found in the heart of QBD, we begin to comprehend never-before imagined possibilities for improving financial reporting and management. By harnessing the potential of this multidisciplinary approach, businesses can revolutionize their financial processes, fostering a new era of efficiency, accuracy, and transparency that benefits all stakeholders and elevates the world of accounting to new heights.

There is a grammar exercise that becomes helpful to understand and use during moments where you are considering previously unexplored overlaps between seemingly disparate disciplines. Imagine you're at the junction of two disparate fields, peering into the chasm where they overlap. You notice a single word, spelled the same in both disciplines yet bearing seemingly distinct meanings. This linguistic phenomenon is known as 'polysemy,' and it serves as a valuable tool when exploring uncharted territories of interdisciplinary overlap. Could these 'polysemic' words, in reality, share aspects of the same definition?

Take, for instance, the concept of 'charge.' In chemistry, 'charges' - positive, negative, and neutral - dictate the behavior of elements. Simultaneously, in the realm of double-entry bookkeeping, we encounter pluses and minuses, debits and credits, that also generate 'charges.' Yet, these 'charges' are rarely considered predictive, certainly not in the same sense as their chemical counterparts. Yet, now that we have observed it, might they be considered as predictive? Intriguing, isn't it? Here, we invite you to journey with us into this intersection, where chemistry's natural law meets the structured logic of accounting, and explore the potential of these shared "charges."

Polarity in debits and credits: In chemistry, polarity refers to the unequal distribution of electrons within a molecule, resulting in regions with partial positive and negative charges. These polar molecules interact with other molecules through attractive forces, such as hydrogen bonding. In accounting, debits and credits represent the two sides of a financial transaction, with debits indicating an increase in assets (or a decrease in liabilities and equity), and credits representing a decrease in assets (or an increase in liabilities and equity). The concept of debit and credit polarity suggests that these opposing forces in financial transactions are inherently interconnected, much like the attractive forces in polar molecules. By viewing debits and credits through the lens of polarity, accountants can explore new ways to understand and analyze the relationships between financial transactions, potentially leading to innovative approaches to managing and optimizing business finances.

QBD can help by enabling teams to test the automated application of polarity to a ledger using a systematic approach to creating Origins and Targets for entries based on this new-found polarity. By creating Targets for credits that meet certain parameters and establishing Origins for certain types of debits, teams can observe and measure the attraction between these elements over time. By adjusting settings on both Targets and Origins, organizations can experiment with different configurations to optimize financial outcomes and gain new insights into the interconnected nature of financial transactions just like chemistry has done with basic elements for decades.

As we implement QBD in ever-broadening circles, it has become apparent that such applications of natural laws of physics can just as easily be applied to accounting theory, the success of which is all but secured through the artful use of QBD in the ledger, itself!

Don't believe me? Consider only this handful of remarkable examples.

Conservation of mass and financial balancing: By applying the QBD principle of ensuring all components are accounted for, teams can create a robust system to ensure that financial statements are balanced, just as the conservation of mass principle is applied in chemistry. By carefully considering the inputs and outputs of a financial system, QBD can help organizations maintain accurate records and ensure financial integrity.

QBD can help by providing a systematic framework for identifying and tracking all components involved in financial transactions. By using QBD tools to analyze the relationships between different financial inputs and outputs, organizations can ensure that their financial statements remain balanced and that no components are overlooked or misrepresented.

Reaction rates and financial forecasting: Implementing QBD can help teams identify the factors that influence the speed of financial transactions and their impact on the business. By using QBD methodologies to analyze and optimize these factors, organizations can improve their financial forecasting and decision-making capabilities, ensuring they react quickly to changes in the business environment.

QBD can help by offering a data-driven approach to identify the factors that influence the speed of financial transactions and their impact on the organization. By leveraging QBD analytics and predictive modeling techniques, teams can assess the potential outcomes of different financial scenarios and make more informed decisions about resource allocation and strategic planning.

Catalysts and process improvements: Teams can leverage QBD principles to identify potential catalysts in their accounting processes, which can speed up or improve the efficiency of financial operations. By implementing targeted process improvements based on QBD insights, organizations can reduce bottlenecks, increase productivity, and optimize their overall financial performance.

QBD can help by providing a methodical approach to identify potential catalysts in accounting processes that can speed up or improve efficiency. By using QBD tools to analyze and optimize these factors, organizations can implement targeted process improvements that reduce bottlenecks and increase productivity.

Equilibrium and financial stability: QBD can help teams identify the factors that contribute to financial stability within their organization, just as chemists study the equilibrium of chemical reactions. By implementing QBD-guided strategies, organizations can work to maintain financial stability and react appropriately to changes in the business environment, ensuring long-term success.

QBD can help by offering a structured approach to identify the factors that contribute to financial stability within an organization. By using QBD methodologies to analyze the interactions between these factors and implement data-driven strategies, organizations can work to maintain financial stability and adapt effectively to changes in the business environment.

Chemical bonds and business relationships: By applying QBD principles to understand the interdependencies between different parts of an organization, teams can better manage and strengthen business relationships.

QBD can help identify the key factors that contribute to strong, mutually beneficial relationships between departments, suppliers, and customers, ultimately leading to more efficient and effective collaboration.

Chemical bonds and business relationships: QBD can help by providing a framework for understanding the interdependencies between different parts of an organization. By leveraging QBD tools to analyze the key factors that contribute to strong business relationships, teams can develop targeted strategies for strengthening collaboration and ensuring more efficient and effective partnerships.

As you can now imagine, the list goes on.

But there is more to this than some seriously cool new accounting theory; there are also seriously solid benefits awaiting those who harness the power of QBD and GAAP integration. By embracing this innovative approach, businesses can enjoy improved decision-making, enhanced transparency and trustworthiness, reduced errors and fraud, streamlined processes, and better regulatory compliance. These advantages set the stage for an exciting new chapter in financial management, where collaboration between financial professionals and QBD practitioners can overcome potential challenges, such as revenue recognition and timing issues, ensuring proper documentation and record-keeping. Through insightful trial and examination made possible through QBD, we will explore the transformative potential of integrating Natural Law with GAAP accounting and learn valuable lessons and best practices from businesses successfully combining these two disciplines.

3. Potential Challenges and Solutions

As businesses embark on the journey of integrating QBD with GAAP accounting, they may encounter a number of challenges. However, with a proactive approach and effective collaboration, these hurdles can be transformed into opportunities for growth and improvement.

A. Revenue Recognition and Timing Issues

One potential challenge in integrating QBD and GAAP accounting is managing revenue recognition and timing issues. The complexities of revenue recognition can be further complicated by the introduction of QBD concepts, especially when it comes to aligning performance obligations with the delivery of goods or services. To address this issue, businesses must ensure that their QBD and accounting teams work closely together to develop a robust revenue recognition framework that accounts for QBD principles. This may involve regular communication, training, and ongoing evaluation of revenue recognition policies to ensure compliance and accuracy.

B. Ensuring Proper Documentation and Record-Keeping

As with any financial system, proper documentation and record-keeping are crucial for the successful integration of QBD and GAAP accounting. Maintaining accurate records of transactions, QBD processes, and compliance efforts is essential for auditing purposes and maintaining transparency. To overcome this challenge, businesses should implement well-defined documentation and record-keeping processes, including the use of standardized templates and centralized storage systems. Regular audits and reviews can help ensure that documentation and records are complete, accurate, and up-to-date.

C. Collaboration between Financial Professionals and QBD Practitioners

The integration of QBD and GAAP accounting requires a high level of collaboration between financial professionals and QBD practitioners. Both parties must understand each other's roles, responsibilities, and expectations in order to ensure a smooth implementation process. This can be achieved through cross-functional training sessions, regular meetings, and open channels of communication. By fostering a culture of collaboration, businesses can ensure that QBD and GAAP accounting work together to drive more accurate and reliable financial reporting and decision-making.

While there may be challenges in integrating QBD and GAAP accounting, businesses can overcome these obstacles by addressing revenue recognition and timing issues, ensuring proper documentation and record-keeping, and strengthening collaboration between financial professionals and QBD practitioners. By proactively addressing these challenges, businesses can unlock the full potential of this innovative approach and enjoy the benefits of improved decision-making, transparency, and compliance.

The Resurrection of Philip B. Crosby

1. Introduction

Philip B. Crosby, a young and ambitious quality control VP at ITT Corporation, was eager to share the news of his team's recent success with his boss. He had worked tirelessly to ensure that their rocket launches were executed flawlessly. With a remarkable record of 10 successful launches out of 11 attempts, Crosby was confident that his boss would be pleased with the results.

The day of the meeting arrived, and Crosby's excitement grew. He had meticulously prepared a report highlighting the team's achievements, envisioning a hearty congratulations and perhaps even a promotion. After a long flight to the company's headquarters, Crosby walked into his boss's office, a spring in his step.

As he presented the report, Crosby's boss listened carefully, nodding as he absorbed the details. When Crosby finished, there was a brief moment of silence, and then his boss spoke. Instead of the expected praise, the words that came out were a challenge: "Somewhere in the world, there's a VP who can give me 11 out of 11 successful rocket launches. I hope it's you."

Crosby was taken aback. In an instant, his expectations were shattered. The long flight home gave him time to reflect on his boss's words, and he realized that the pursuit of quality had no finish line. It was not about just meeting targets, but rather about striving for continuous improvement and pushing the boundaries of what was possible.

As the plane cruised through the sky, Crosby pondered what he could do differently. He began to envision a new approach to quality management, one that would involve everyone in the organization and engender a culture where quality was an integral part of the company's DNA.

But how to pay for it?

The inspiration from that one thought became a philosophy to be captured in several books like lightening in a bottle, most notably "Quality is Free," which used a very clever accounting trick that would forever change the way businesses approach quality as a matter of double-entry bookkeeping, or so we thought until the 1996 GAAP-induced confusion.

As Crosby contemplated his new approach to quality management, he developed an innovative strategy to demonstrate the financial benefits of investing in quality. He created a ledger entry called "hassle," which represented the costs associated with each defect. By quantifying these costs, Crosby could justify investments in quality improvement initiatives.

He tasked every employee with finding hassles in everything they experienced – even setting up reward frameworks to promote the effort.

Then, his method involved allocating capital to reduce the hassle and subsequently measuring the returns on investment. When the decrease in hassle costs exceeded the amount invested, Crosby could show tangible profits generated from the quality improvement efforts. This groundbreaking approach not only highlighted the importance of quality but also provided a financial framework to support continuing quality initiatives, bridging the gap between quality management and business profitability.

By the way, he reached the goal of 11 out of 11 successful launches the next year and turned a profit while doing it.

Crosby's innovative approach to quality management and its financial implications quickly gained traction in the business world. His philosophy resonated with organizations seeking to improve their performance, streamline operations, and boost profitability. As a result, Crosby's ideas spread like wildfire across industries and around the globe.

Many major corporations adopted his methods, recognizing the potential for significant financial and operational improvements. One notable example was IBM, which embraced Crosby's principles and established the IBM Quality College. The Quality College aimed to educate and train IBM employees in Crosby's quality management techniques, ensuring that the entire organization was aligned with his philosophy.

This widespread adoption of Crosby's approach solidified his methods as a standard in quality management. His work transformed the way businesses perceived and pursued quality, leaving an indelible mark on global business practices and shaping the future of quality management.

But then what happened? Why is it that the approach is all but disappeared today?

While GAAP has been evolving for many decades, it is true that there were significant changes in accounting principles and practices in the mid-1990s—specifically around 1996. Unfortunately, this period also coincided with a growing misinterpretation of GAAP's intentions, particularly regarding the concept of **assessing the useful life of assets, calculating bad debt allowances, and valuing intangible assets like intellectual property** *(mentioned in Fall Into the GAAP earlier)*, and more specifically, Intellectual Viability. The wrongheaded and misinformed notion began to spread through corporate leadership ranks that until a project reaches milestone 3, which is where it may be declared to have reached the tipping point of Intellectual Viability, all spending on that project must be considered expense only—which would preclude Crosby's clever account lever. As a result, many executives were led to believe that investing in quality until the end of a project was no longer viable under the new accounting standards. Oh sure, it was "best practice," but it was also a dead expense – something to minimize until later in the life cycle, where admittedly the remedy of hassle costs far more to accomplish.

Gone was the axiom, "The earlier in the life cycle that a defect is found, the cheaper it is to fix."

Philip B. Crosby, a pioneer in quality management, indeed lived through this period of confusion and misunderstanding, witnessing the decline in the application of his quality concepts. He passed away in 2001, leaving behind a legacy that was, at the time, overshadowed by misconceptions and misinterpretations of GAAP.

It is indeed a tragic story of timing and misunderstanding, where a visionary leader in quality management lived to see his groundbreaking ideas fade from the forefront of business thinking. It serves as a cautionary tale about the importance of proper education and interpretation of accounting standards and the potential consequences of failing to recognize the true value of investing in quality.

But this story ain't over yet…

2. Implementing Quality in a GAAP Compliant Way

As people start to implement Quantum Business Dynamics™ (QBD™) to accounting, they are beginning to understand more about GAAP, and the misunderstandings around things like Intellectual Viability. The misconception that GAAP prevents businesses from investing in quality improvements has persisted for years, mainly due to a lack of clarity regarding the accounting treatment of such investments. However, a closer examination of GAAP guidelines reveals that it is indeed possible for companies to implement Crosby's approach while remaining compliant with GAAP.

In terms of Intellectual Viability, the misunderstanding that it somehow prohibits investing in quality improvements is unfounded. Intellectual Viability refers to the ongoing economic value of a project, such as a software development initiative. However, this concept does not preclude organizations from investing in quality improvements to address existing defects or hassles.

One of the key components of Crosby's approach is the concept of "hassle" – the costs associated with defects across the company. Under GAAP, companies can recognize these costs as expenses on their financial statements, as long as they meet specific criteria. For example, ASC 420-10-25-13 provides guidance on recognizing liabilities associated with exit or disposal activities, including the cost of addressing product defects.

To invest in quality improvements, companies can capitalize on investments that result in future economic benefits. As long as the investment meets the criteria for capitalization under GAAP, such as ASC 350-40 for internal-use software, it can be legitimately accounted for on the company's financial statements.

By understanding the nuances of GAAP and the underlying principles of Crosby's quality approach, companies can adopt the Crosby method today without violating GAAP. Quantum Business Dynamics™ (QBD™) can play a pivotal role in helping businesses better understand the relationships between ASC 420-10-25-13 and ASC 350-40. By leveraging the principles of QBD, companies can create Origins and Targets that link the guidelines of these accounting standards with their financial ledgers.

QBD's inherent ability to provide a clear and transparent view of the connections between various financial elements can be profoundly important when it comes to explaining these relationships to regulators. With the use of QBD, businesses can ensure that their accounting practices are not only GAAP-compliant but also easily auditable and traceable. This level of transparency, combined with the immutable nonrepudiation offered by QBD, can provide a solid foundation for defending the company's accounting methods should any questions be raised by regulators.

This newfound understanding of GAAP compliance, coupled with the power of QBD, has the potential to re-ignite the healthy use of Crosby's quality methods in the business world. It opens up significant opportunities for large firms, such as The Big Five, to reintroduce approaches like "Quality is Free" for the betterment of businesses across all industries. By combining the insights of QBD with the principles of Crosby's quality approach, businesses can unlock new levels of efficiency, accuracy, and transparency in their financial reporting, leading to improved decision-making and overall operational excellence.

By understanding the nuances of GAAP and the underlying principles of Crosby's quality approach, companies can adopt the Crosby method today without violating GAAP. Executives who may have been educated in 1996 on GAAP and never revisited their understanding could benefit from a thorough review of current GAAP guidelines and a more in-depth exploration of the possibilities for integrating quality investments into their financial accounting practices.

By dispelling misconceptions around GAAP and Intellectual Viability, businesses can embrace the power of Crosby's quality methods alongside QBD, enhancing their financial performance and overall operational excellence.

But what would it all look like? How would applying Natural Law to double-entry bookkeeping, as explored in the last section, impact this discussion of the Crosby method on the actual ledger?

The Binary Link Chain approach offers a novel way for companies to manage their resources, investments, and quality efforts by applying QBD to their accounting practices. The design pattern consists of several steps that guide a company in renewing quality approaches, such as Crosby's method:

A. In QBD, think of debits (+) as Origins. These are positively charged and act like actionable validations, or Sources, that are drawn to Targets, which are negatively charged (-).
B. When a hassle occurs in the field, it's analyzed and the associated cost is confirmed by an expert. This validated hassle becomes a Source (+), ready for action.
C. A broker—someone involved in quality efforts—connects this hassle Source (+) to a Target (-). This Target could be a goal of identifying a certain number of hassles in a set period within an organization. This connection creates a QBD Binary Link (+), representing a value that is added to the ledger as a forecasted and budgeted hassle.
D. When this hassle becomes a Binary Link (+) on the ledger, it's positively charged. Keep in mind that Binary Links also serve as Origins, just like Sources. This means that they can naturally connect to various Targets (-), such as a plan to invest in quality improvements aimed at reducing similar hassles.
E. When these hassle Binary Links on the ledger are connected to a quality investment Target, the company saves money by avoiding forecasted hassles. This avoidance creates individual Binary Links of profit (+).
F. These profits (+), now positively charged as Binary Links, can naturally connect to other Targets (-). These could include planned investments, dividends for stockholders, or employee benefits. This process creates a cycle of continual improvement and value generation.

This illustration vividly captures the story of how positively charged elements are naturally attracted to and link to negatively charged ones, resulting in possessable value. This value can then further be attracted to more Targets. Intriguingly, this lends more credence to pondering the centuries-old polysemies that exist between chemistry and accounting, such as the term "charge."

As companies adopt this Binary Link chain methodology, and in collaboration with financial professionals, auditors, and regulators, they can confidently weave this innovative concept into their accounting fabric. Doing so enables them to leverage Quantum Business Dynamics™ (QBD™) to breathe new life into Crosby's quality methods and enhance their financial performance.

3. Conclusion

In conclusion, the powerful synergies between Quantum Business Dynamics™ (QBD™) and Crosby's "Quality is Free" theory offer businesses an innovative and effective way to revitalize their quality management efforts. By combining the principles of QBD with Crosby's methods, organizations can unlock new avenues for growth, efficiency, and profitability. The potential benefits of this approach are vast, including enhanced decision-making, improved transparency and trustworthiness, reduced errors and fraud, streamlined processes, and better regulatory compliance.

As we continue to explore the exciting frontier of integrating QBD and Crosby's quality theory, there are boundless opportunities for further research and implementation. Companies and researchers alike should embrace the challenge of discovering new ways to leverage the strengths of both fields, ultimately driving the evolution of business practices towards a more collaborative, effective, and profitable future.

The marriage of QBD and Crosby's methods holds the promise of transformative change in the world of accounting and quality management. By nurturing a deeper understanding of the underlying principles, and by encouraging collaboration between financial professionals, QBD practitioners, and quality experts, we can create a new era of innovative solutions that benefit businesses, stakeholders, and society as a whole. Let us move forward together in the pursuit of this remarkable potential, and shape the future of accounting and quality management for the better.

Decentralized and Segmented Payments: Disrupting Chains, Creating Links

Maria lives in a remote village in Nicaragua. She has a small farm where she grows coffee beans, but she struggles to make ends meet. Maria has always wanted to sell her coffee directly to consumers in the United States, but the lack of financial infrastructure in her village makes it nearly impossible for her to receive payments from abroad.

Across the globe in New York, Lisa is a passionate advocate for fair trade and sustainable farming practices. She's discovered Maria's coffee through a mutual contact and is eager to source it for her newly opened coffee shop. However, Lisa is frustrated by the high fees and long delays associated with international wire transfers. The opaque nature of these transactions leaves her uncertain about how much of her payment is actually reaching Maria.

Meanwhile, in San Francisco, Tom is a software developer with a knack for blockchain technology. He believes in its potential to democratize finance and has been developing a decentralized payment app that could solve the problems Maria and Lisa face.

Tom's app utilizes blockchain technology to make fast, secure, and low-cost international payments. It's a peer-to-peer system, which means that Lisa can pay Maria directly, without intermediaries. The app also supports segmented payments, which allows Lisa to break down her payments into smaller parts. This is especially useful when she needs to pay multiple coffee growers at once.

Through a mutual contact, Lisa and Maria connect with Tom. They decide to trial his app. Lisa is thrilled to see that her payments are reaching Maria almost instantly, with significantly reduced costs. Maria, on her end, is able to receive funds on her basic smartphone, and for the first time, she feels included in the global economy. The transparency of the blockchain ensures that Lisa knows her payments are directly impacting Maria's life, growing trust between them.

The story illustrates the transformative potential of decentralized and segmented payments. However, it's important to acknowledge that challenges still exist, including regulatory issues, scalability, and the need for digital literacy. Despite these challenges, the promise of a more inclusive and efficient financial system offers a compelling vision for the future.

But things are not always sunshine and roses …

Jacob is a part of a notorious cybercriminal group based in an undisclosed location. He specializes in ransomware attacks, where he encrypts victims' data and demands payment for the decryption key. The group targets businesses, hospitals, and sometimes even cities, causing immense damage and disruption.

Jacob's group used to demand payment via wire transfer, but that left a trail that law enforcement could follow. With the advent of decentralized payments, they've found an effective method to anonymize their transactions. They now demand payment in cryptocurrencies, making it much harder for authorities to trace their activities.

In a recent attack, they targeted your hospital network, crippling its operations. Unable to access patient records, the hospital was forced to divert critical care patients to other facilities. The group demanded a significant sum in cryptocurrency, which due to the segmentation feature, was to be paid to various digital wallets, further muddying the trail.

The hospital had no choice but to pay the ransom, as lives, like your spouse's, were at stake. The segmented payments were quickly dispersed into various wallets, which were then drained into cryptocurrency mixers for additional obfuscation. As a result, the cybercriminals were able to walk away with their ill-gotten gains, leaving the hospital network to pick up the pieces.

This darker scenario is a real concern for law enforcement and regulators globally. Decentralized and segmented payments can indeed be a tool for good, but they can also be exploited by malicious actors. It's not just about maintaining power, but also about ensuring the safety and security of society. Therefore, as we navigate this brave new world of decentralized finance, it's crucial to find a balance between innovation, freedom, and security.

This might shed light on why the stalwarts of centralized finance are becoming more vocal, clamoring for stricter regulations that could potentially fetter the blossoming decentralized finance sector (DeFi) with the same rules they've been bound by, such as the travel rule. Are these institutions acting out of selfless concern for societal welfare, or are they merely seeking to protect their own stronghold, sensing its potential erosion? Could Quantum Business Dynamics™ (QBD™) provide solutions to these quandaries? To answer these questions, it may be beneficial to first delve into some underlying principles.

Decentralized Payments:

Imagine you want to send money to your friend who lives in another city. Traditionally, you'd use a bank or a service like PayPal to transfer the money. These services act as middlemen, taking your money, checking the details, and then sending it to your friend's account. This is a centralized system because one entity (the bank or PayPal) has control over the entire transaction.

Now, imagine if you could send the money directly to your friend without needing a bank or PayPal. Instead, a network of computers spread across the world verifies the transaction. This is what decentralized payment is. It's like a group of friends passing a note in a classroom. The note goes directly from you to your friend, with everyone in the room (the network of computers) confirming that the note was passed correctly.

Segmented Payments:

Now, let's say you owe your friend $100 but you only have $10 right now. You promise to pay the remaining $90 later. This would be an example of a segmented payment. You're breaking the full payment into smaller 'segments' that you pay over time.

Decentralized payments and segmented payments can work together. For example, you could use a decentralized payment system to send segmented payments. It's like passing your friend a series of notes (segmented payments) in class (decentralized system) until you've sent the full message.

In both cases, the key concept is about making payments easier, more flexible, and less dependent on a single authority like a bank. However, they solve different aspects of the payment process: decentralization is about who processes the payment, while segmentation is about how the payment is divided.

The "Travel Rule" is a term that comes from banking and financial regulations. Imagine you're sending a letter to a friend in another city. You wouldn't just write the letter and send it off. You'd also have to include your return address and your friend's address on the envelope. This way, if there's any problem with the letter, or if anyone needs to know where it came from or where it's going, that information is readily available.

In the financial world, the "Travel Rule" works similarly. It's a regulation that requires certain information to "travel" along with financial transactions over a certain amount. This information typically includes the names and account numbers of the sender and receiver, and other identifying information.

The rule was created to prevent money laundering, terrorist financing, and other illegal activities. It provides a way for law enforcement to track funds and investigate if anything suspicious is happening.

In the context of decentralized or crypto payments, the "Travel Rule" presents challenges because these types of transactions were designed to be more private and less tied to personal information than traditional banking. In fact, a crucial component of the Travel Rule is that it mandates the transmission of specific sender and recipient information to accompany every qualifying fund transfer. This notification must occur concurrently or ahead of the transaction, a requirement that poses a significant challenge for many existing decentralized systems—especially when they weren't architected to support such communications. However, regulators around the world are increasingly expecting compliance with rules like this to ensure that these new financial systems aren't being used for illegal activities.

There are other regulations that are routinely brought up in the DeFi space:

- AML (Anti-Money Laundering) Regulations: These rules are designed to help detect and report suspicious activity including the predicate offenses to money laundering and terrorist financing, such as securities fraud and market manipulation.

- KYC (Know Your Customer) Regulations: These regulations require financial institutions to verify the identity of their clients. They are a critical component of AML policies.

- CFT (Combating the Financing of Terrorism) Regulations: CFT regulations aim to prevent and detect the movement of funds intended for use in acts of terrorism.

- FATF (Financial Action Task Force) Guidelines: The FATF is an inter-governmental body that sets standards to promote effective legal, regulatory, and operational measures to combat money laundering, terrorist financing, and other related threats.

- Securities Regulations (e.g., SEC Regulations in the U.S.): These laws are designed to protect investors; maintain fair, orderly, and efficient markets; and facilitate capital formation. They can apply to certain types of cryptocurrencies and initial coin offerings (ICOs).

- GDPR (General Data Protection Regulation) in the EU: This law protects individuals' privacy and sets guidelines for the collection and processing of personal information within the EU. It's relevance comes in when considering privacy coins or other privacy measures within the crypto space.

- The Bank Secrecy Act (BSA): This U.S. law requires financial institutions to assist U.S. government agencies in detecting and preventing money laundering.

- MiFID II (Markets in Financial Instruments Directive II): This is a legislative framework instituted by the European Union (EU) to regulate financial markets in the bloc—a group of European countries—and improve protections for investors.

These regulations and others like them pose significant challenges to the decentralized and often anonymous nature of cryptocurrency transactions. However, they also underline the importance of building systems like QBD that can provide the needed transparency and traceability while still preserving the benefits of decentralized finance.

Maria, our promising entrepreneur from a small town in Nicaragua, could face several concerns due to the "Travel Rule".

1. Privacy: Maria relies on the privacy that her decentralized payment system affords her. In a situation where her transactions need to comply with the "Travel Rule", she might worry about her personal and financial

information being exposed to dangerous elements in her country. This could potentially put her at risk.

2. Complexity: Complying with the "Travel Rule" can be technically complex. It requires integrating systems that can capture and transmit the necessary information alongside transactions. For Maria, who's not a technical expert, this could pose a significant challenge.

3. Cost: Compliance often comes with a cost. From developing or purchasing compliant software systems to possibly needing to hire a professional to ensure she's in line with the rule, these are costs that Maria, as a small business owner, may not be equipped to handle.

4. Regulatory Risk: If Maria were to inadvertently fail to comply with the "Travel Rule", she could face penalties, which could harm her business financially and potentially lead to the loss of her ability to make transactions.

5. International Transactions: If Maria wants to expand her business and start transacting with international customers or suppliers, the "Travel Rule" could become even more complex due to differences in regulations between countries. This could limit her growth potential.

The "Travel Rule" is indeed essential for preventing financial crimes, but for honest individuals like Maria who use decentralized or segmented payments for legitimate reasons, it can present substantial challenges.

Meanwhile, Jacob, the devious individual from our previous example, would also have his concerns with the "Travel Rule," although for entirely different reasons than Maria.

1. Anonymity: Jacob values the anonymity that decentralized payment systems provide, which allows him to carry out illicit activities without being easily traced. The "Travel Rule" would require Jacob's personal information and details about his transactions to be shared with the authorities, potentially exposing his illicit activities.

2. Legal Consequences: If the "Travel Rule" were to expose Jacob's unlawful activities, he could face significant legal consequences, including fines, asset seizure, or even imprisonment.

3. Difficulty in Finding Partners: As the "Travel Rule" is enforced, those who wish to remain compliant will avoid dealing with parties who refuse to adhere to the rule. This could make it harder for Jacob to find willing partners for his illegal transactions.

4. Complexity and Cost: Just like Maria, Jacob would also face the technical complexity and costs associated with complying with the "Travel Rule". However, his motivation would be to find ways to bypass these controls rather than to comply with them.

In short, while the "Travel Rule" poses challenges for legitimate users like Maria, it's designed to disrupt the activities of individuals like Jacob, who seek to exploit the financial system for illicit purposes.

In this complex arena of interests, Quantum Business Dynamics™ (QBD™) could play a substantial role in satisfying the needs of the varying parties.

1. Maria and the Good People: QBD, with its model of Binary Links, could facilitate the secure, efficient and affordable execution of decentralized and segmented payments. This would bring financial services to unbanked or underbanked individuals like Maria. The Binary Links would ensure that transactions are transparent and can be tracked, while providing a level of security, privacy, and trust in the system.
2. Jacob and the Bad People: The transparency and traceability of Binary Links would make it harder for illicit activities to go undetected. This is because every transaction leaves traces that can be evaluated with proper legal authorization, and this could deter individuals like Jacob from using the system for illegal purposes.
3. Technologists like Tom: QBD would be a boon to technologists as it provides a new, innovative framework for financial transactions, thus expanding the possibilities of what can be achieved with decentralized technologies. QBD could provide new challenges and opportunities for innovation.
4. Bankers: Rather than seeing DeFi as a threat, QBD could help banks identify new roles and opportunities for growth. Banks could facilitate virtual safety deposit boxes, helping people in the DeFi markets with decentralized recovery (like forgetting your password – a real issue in defi, but skill banks, but skill banks, institutions known for their specialized expertise, excel at). They could also look at extending forms of insurance by reclassifying some elements of deposits—a very innovative concept that would likely further have a stabilizing effect on crypto markets. This could open up new revenue streams and growth opportunities, helping them transition into the new financial era.
5. Regulators and Governments: QBD's inherent transparency could assist regulators in monitoring transactions, reducing the cost of oversight and potentially reducing the need for additional revenue through fines. The traceability of Binary Links could assist in the enforcement of regulations like the "Travel Rule", making the whole system more robust and less prone to misuse.

It is true that many decentralized finance (DeFi) platforms and enthusiasts are indeed resistant to regulatory changes that they perceive as threatening the fundamental principles of decentralization, privacy, and autonomy upon which these systems were built. They argue that imposing traditional banking regulations, like the Travel Rule, onto decentralized platforms could fundamentally alter their architecture, making them less efficient, less private, and thus less desirable for users.

However, it's crucial to recognize that the growing adoption of DeFi platforms and cryptocurrencies comes with increased scrutiny from regulators worldwide. The challenge lies in finding a balance between preserving the innovative, decentralized nature of these platforms while also ensuring they're not used for illicit activities.

That's where Quantum Business Dynamics™ (QBD™) could come in. QBD, with its flexible and interoperable structure, could potentially act as a bridge between DeFi platforms and regulatory requirements. QBD's asynchronous interfaces could be implemented at key moments during transactions on DeFi platforms, allowing for regulatory compliance without disrupting the core architecture of these systems.

By providing an alternative communication band, QBD could allow for the necessary transfer of information to meet regulatory requirements, such as the Travel Rule. This wouldn't necessitate a redesign of DeFi platforms but rather provides a parallel solution that facilitates regulatory compliance while preserving the inherent advantages of decentralization.

So, in essence, QBD could potentially alleviate some of the concerns expressed by DeFi platforms about altering their systems to comply with regulations. However, a wider discussion and detailed analysis would be required to understand the full implications and feasibility of such an integration.

One last note on communication is important. QBD is not a transport. What does that mean? In telephony, the term "transport" refers to the infrastructure or medium that carries voice, data, or video from one place to another. It's the pathway, the road that data travels on.

In the context of QBD, it's crucial to understand that QBD does not act as this pathway or road. Instead, it's more like a versatile traffic director that can interact with multiple roads or transport layers. This means that QBD can trigger and respond to messaging events, regardless of the transport layer used.

Imagine a bustling city with various types of transport - buses, trams, cars, bicycles, pedestrians. QBD is not the road or the tram tracks or the pedestrian walkways. Instead, it's the traffic management system that ensures everyone can get where they're going as efficiently and safely as possible, irrespective of their mode of transport.

Now, let's consider how this can apply to the Travel Rule and DLT. Currently, some DLTs lack a built-in transport layer for the type of information required by the Travel Rule. But with QBD's interoperable nature, it can work alongside various transport layers - think of it as a traffic management system that can work with any type of road.

An existing and widely used transport layer, such as HTTP, SMTP or even a blockchain-based transport like Whisper, could serve as the road for the information. QBD, in its role as traffic manager, could then ensure that the necessary information is sent and received at the right times, helping the DLT comply with the Travel Rule without needing to build a whole new transport infrastructure.

This approach not only eases the burden on individual players but also cultivates a more collaborative, efficient ecosystem. Each player can focus on their core competencies while contributing to the larger solution. In the same way that a city's transport system works best when each element - roads, traffic lights, signages - does its job effectively, this federated approach allows for a more robust and adaptable system.

Banks, traditionally burdened with compliance duties, could also benefit from this shift. As they transition to more decentralized models, adopting an interoperable system like QBD could future-proof their operations. They could more easily adapt to changing regulations, and even preemptively prepare for future requirements, by federating functionality to different entities within the system. In essence, they could become a cog in a larger, more resilient machine, rather than shouldering all the responsibility themselves.

Overall, QBD's application in the realm of decentralized and segmented payments could provide a balance between the freedom and innovation of DeFi and the need for security, transparency, and regulatory compliance. It has the potential to create a win-win situation for all parties involved, promoting economic inclusion, deterring illicit activities, enabling technological innovation, providing new opportunities for traditional finance, and aiding regulatory oversight.

Data Janitor in a Drum

It's an open secret that in today's data-driven world, the quality of corporate decision-making is only as good as the data that fuels it. Yet, a startling number of businesses grapple with the incessant plague of subpar data quality — a malady that not only undermines their operational efficiency but also corrodes the foundation of strategic planning and forecasting.

It's not an exaggeration to say that the quality of data can make or break a company's future. Yet, data quality issues persist, becoming more than a mere nuisance — they are a fundamental hindrance to business growth, agility, and innovation.

Disparate Systems and Formats: Many organizations are awash with data from various Sources, each with its unique format and structure. Integrating these disparate data systems into a cohesive whole is a colossal task, often leading to jumbled, inconsistent, and unreliable data.

Inadequate Data Governance: Without a robust data governance framework, maintaining data integrity becomes an uphill battle. The lack of clear policies and procedures for data management often results in data inaccuracies, duplication, and non-compliance with regulations like GDPR, HIPAA, or the Travel Rule.

Rapid Data Decay: Data, especially customer-related data, decays at an alarming rate. Contact details change, customers relocate, businesses shut down — and if this changing data isn't updated promptly, it leads to an erosion of data quality.

Human Error: Humans, by nature, are prone to mistakes, and these errors creep into data. Whether it's due to typos, misinterpretation, or negligence, human error is a persistent contributor to data quality issues.

Poor Data Literacy: Many organizations lack a workforce with strong data literacy skills, leading to misinterpretation and misuse of data. This not only impacts decision-making but also impairs the organization's ability to maintain data quality.

In the context of Natural Law, these challenges are akin to the chaos found in nature. But just as nature finds a way to create order and systems, so too can businesses with the right approach. By leveraging principles from Accounting, Quality Management, and Regulations, businesses can bring order to the chaos. Quantum Business Dynamics™ (QBD™) offers a way to navigate this labyrinth, providing businesses a path to higher data quality, better decision-making, and ultimately, improved financial performance.

In grappling with their data, Chief Data Officers (CDOs) have long relied on traditional data management approaches. Yet these approaches, while foundational, have significant limitations that make them less effective in today's complex and rapidly changing data landscape. Here are some of the most critical limitations:

1. Scalability: Traditional data management tools often struggle to scale to meet the demands of today's big data environment. They were not designed to handle the volume, variety, and velocity of data that businesses now produce and collect.

2. Siloed Data: Traditional data management practices often lead to data silos — information trapped within a department or a specific software application, inaccessible to other parts of the business. This fragmentation can lead to inconsistencies, redundancies, and an incomplete picture of the business.

3. Data Integration Challenges: With data coming from diverse sources, in various formats, integrating this data into a unified, coherent system is a significant challenge. Traditional approaches often lack the flexibility to handle this complexity effectively.

4. Data Quality: Ensuring data accuracy, consistency, and reliability is a perennial challenge. Traditional data management practices often lack robust data quality checks, leading to errors and inaccuracies creeping into the data over time.

5. Real-Time Insights: Traditional data management practices often involve batch processing, which can delay the availability of insights. In today's fast-paced business environment, this lag can mean missed opportunities.

6. Data Governance and Compliance: Implementing effective data governance and ensuring compliance with an array of data privacy and protection regulations is a significant challenge. Traditional approaches often lack the robustness to ensure consistent adherence to governance policies and regulatory requirements.

7. Data Security: Protecting data from breaches, theft, and loss is a critical concern. Traditional data management systems often have vulnerabilities that can be exploited by malicious actors.

In a nutshell, while traditional data management approaches provide a starting point, they are not sufficient to meet the challenges of today's complex and high-stakes data environment.

A letter from every CDO ever:

Dear CFO,

In the digital age, data has become the lifeblood of our business, driving strategic decisions and fostering innovation. Over the years, our Chief Data Office (CDO) has been striving to harness the full potential of our data assets. However, traditional data management approaches, while serving us well in the past, are no longer sufficient to meet our current and future needs. This is not just a technological issue, but a strategic one that impacts our bottom line, and here's why.

Firstly, scalability has become a major concern. Our traditional systems struggle to handle the ever-growing volume, variety, and velocity of data we now generate and collect. This limitation impedes our ability to fully leverage big data, leading to missed opportunities and a potential competitive disadvantage.

Secondly, data silos are causing inefficiencies and limiting our ability to get a comprehensive, 360-degree view of our business. Data trapped within specific departments or software applications cannot easily be accessed or utilized by other parts of the business. This fragmentation results in inconsistencies, redundancies, and a lack of holistic understanding, which can lead to suboptimal decision-making.

Thirdly, the challenges of integrating diverse data sources and formats are significant. Our traditional data management practices lack the flexibility to effectively handle this complexity. As a result, we're missing out on the synergies that could be realized by integrating our disparate data assets.

Next, ensuring data quality is a persistent challenge. Without robust quality checks, our data may contain errors and inaccuracies, which can lead to misguided decisions and actions.

Furthermore, the delay in getting insights due to batch processing inherent in our traditional systems is a significant issue. In today's fast-paced business environment, real-time insights can mean the difference between seizing a market opportunity and missing it entirely.

Finally, data governance and compliance, as well as data security, are significant challenges. Ensuring consistent adherence to governance policies and regulatory requirements, and protecting our data assets from breaches, theft, and loss are critical. Traditional data management systems often lack the robustness to effectively address these issues.

In summary, the limitations of our current data management approach are hindering our ability to fully leverage our data assets, leading to suboptimal decision-making, missed opportunities, inefficiencies, and potential regulatory and security risks. The need for a more accessible and cost-effective solution has never been greater. As we move forward, it's critical that you provide me with more money so that we can consider new approaches that can address these challenges and help us to truly unlock the potential of our data.

Sincerely,

Your CDO

Building a Parallel Bi-link Fabric:

Quantum Business Dynamics™ (QBD™), as an emerging hybrid standard of centralized and decentralized best practices, empowers your business to achieve its key goals by capturing a comprehensive deconstruction of your existing business practices and systems while leaving your data in place. This state-of-the-art methodology governs automated adapters that seamlessly integrate with your existing data and interfaces, without copying or interrupting them. Through this coherent integration, QBD gleans standardized particles of four different types, each characterized by predictable rules and behaviors. These particles serve as the building blocks to rapidly reconstruct a meta model of your company that evolves with your business over time, while preserving your original practices and systems. This QBD meta model provides a robust foundation for directly applying AI, ML, and other evaluations, effectively eliminating the need for costly data janitor work within your original platform. Additionally, it equips you with strategic levers to cleverly model and optimize your business operations going forward, all while leveraging your existing assets.

Picture if you will, a journey into a different dimension. Not one of sight and sound, but of data. You are about to cross over into... the Bi-link Zone.

Imagine, for a moment, your business - a bustling enterprise with all its complex machinery, the interweaving processes, the ceaseless flow of information. It's a vibrant, yet daunting landscape, filled with data silos, legacy systems, and a constant battle against inefficiencies. This is the world as you know it, the world you've grown accustomed to, as have so many others in your position.

Now, look next door. On an adjacent lot, a new structure is being built. It's a mirror image of your existing business, but it's different. It's gleaming, efficient, streamlined. There's a sense of order, of harmony. This is not some alternate reality or distant future vision; it's a parallel construct of your business, forged in the crucible of a unique concept: the Bi-link Fabric.

In this parallel construct, the characteristics of your business data and processes are mirrored, but they're free of the shackles of your current system. Without copying or moving your data, these characteristics are structured, standardized, and interconnected, forming a grand tapestry of Binary Links. Each piece of data, each process, each decision, is a Binary Link - a building block that fits seamlessly into this new, gleaming edifice.

The Bi-link Fabric is not just a data framework; it's a different way of viewing and managing your business. Each Binary Link is an Origin or a Target, positively or negatively charged, naturally attracting or being attracted to other links. This creates a dynamic, self-organizing system where information about data is not just stored, but actively used and linked in meaningful ways.

But this isn't a simple duplication of your existing business. This parallel construct is constantly learning, growing, evolving. Every interaction, every decision, every new piece of data adds another link to the fabric, making it richer, more intricate, more powerful.

In this Bi-link Zone, the challenges of jumbled data, of siloed systems, of ineffective decision-making are fading echoes of the past. The future is here, and it's not a spooky, alien landscape. It's a familiar world, just... better. More efficient. More insightful. More adaptable.

So, as we step into this Bi-link Zone, let's leave behind our preconceptions, our fears, our old ways of doing things. Let's embrace this new world of possibilities, where data is not a challenge to be overcome, but a resource to be harnessed, a tool to be utilized, a path to a brighter, more efficient, more profitable future.

Welcome to the Bi-link Zone.

The benefits of creating a Bi-link Fabric:

Imagine walking into a bustling market, vendors shouting, goods being exchanged, an overwhelming cacophony of commerce. This is the state of the data within many organizations today: abundant, chaotic, and full of potential. Yet, the key insights needed are often buried deep within the noise. Increasing accessibility to these data insights is not just about sorting through the chaos, but creating a parallel structure that organizes this abundance into a form that speaks your language.

Now, picture this. As your business processes unfold, generating transactions and interactions, a parallel structure is being built. This structure, formed by Binary Links, reflects the agreements and commitments that underlie your business. It's not a duplicate or transformation of your existing data, but rather a sort-of mirror image construct that captures the essence of your operations. The beauty of this approach is that it provides a direct access to the context and meaning of your business activities.

Let's talk about efficiency and cost-effectiveness. The traditional approach of data management often involves a painstaking process of data cleaning, transformation, and normalization. But why dig through the jungle when you can walk on a paved road? With the Bi-link Fabric, you can bypass the time-consuming data janitor work on your legacy data, and instead dive directly into the Bi-link mirror construct for the insights that matter, saving resources that can be better utilized elsewhere.

Lastly, let's consider scalability and adaptability. As your organization grows and evolves, your data landscape will naturally become more complex. But the Bi-link Fabric is not just a static structure; it's a dynamic, adaptable system. As new agreements are made, new interactions occur, the fabric evolves in real-time, reflecting the current state of your business. It adapts and scales with you, ensuring that you're always looking at a true reflection of your operations.

In essence, by interfacing with the systems and processes that feed data into your existing systems, you can instantiate a Bi-link Fabric that provides a real-time, contextual, and accessible view of your business activities. And the best part? Your existing data systems remain unchanged, while you gain an innovative tool to unlock deeper insights.

Democratizing the Process of Managing and Understanding Complex Data Relationships:

Embedding a culture of continuous enhancement and data-informed decision making is akin to laying a robust foundation for a skyscraper. Without it, the structure may still stand, but it's likely to sway precariously with the winds of change. QBD provides a set of tools to lay this foundation, but what does it look like in practice?

First, let's consider the concept of continuous improvement. The traditional approach might involve periodic audits, retrospective analyses, and change management programs. But with QBD, improvement becomes an inherent part of your day-to-day operations. As your employees learn to identify and use the basic Particles in their sphere of influence, they start to see their work through a new lens. They can spot inefficiencies, propose solutions, and observe the impact of their changes in real-time. It's about empowering every member of your team, from the CEO to the security guard, to take ownership of the data and the insights it offers.

This brings us to the concept of data-driven decision making. With traditional data management approaches, data scientists often bear the brunt of the work, spending more time cleaning and organizing data than deriving insights from it. In contrast, QBD democratizes the process, making complex data relationships understandable and accessible to all. The role of data scientists shifts from data janitor to data guide, helping others navigate the Bi-link Fabric and extract meaningful insights.

The impact of this shift can be transformative. Imagine a meeting where everyone comes prepared with data-informed insights, not just the data science team. Imagine a culture where every decision is backed by tangible evidence, not just gut feelings. Imagine an organization where everyone is a steward of data, continuously striving for improvement.

In essence, implementing QBD is more than just adopting a new data management approach. It's about cultivating a culture that values data, embraces improvement, and encourages everyone to play an active role in shaping the organization's future. The Bi-link Fabric is not just a tool; it's a catalyst for cultural transformation.

The journey towards a democratized data culture begins with a paradigm shift in how we view and interact with our data. Rather than seeing data as a by-product of our work that needs to be wrangled into coherence, QBD encourages us to view data as the end product itself. This shift not only simplifies data management but also unlocks new possibilities for empowerment and collaboration.

Picture this. Your data team uses software adapters to create a preliminary Bi-link Fabric from your existing data. This isn't about cleaning or transforming your data but about using it as a steppingstone to a new, more effective approach to data management. This Bi-link Fabric acts as a skeleton, a basic framework upon which your entire organization can build.

But how does the rest of the organization get involved? Let's look at a simple example. Imagine you have a customer service representative named Alice. Alice isn't a data scientist, but she interacts with valuable data every day. With a basic understanding of the four Particles and a user-friendly dashboard, Alice can start creating Sources and Targets based on her daily interactions. Maybe Alice notices a recurring issue with a product, or maybe she identifies a trend in customer feedback. She can capture these observations in the Bi-link Fabric, ensuring that they're not lost in the shuffle of her day-to-day work.

On the other side, your data scientists or data brokers can build Bridges between Alice's Sources and Targets, and those created by others, linking disparate data points into a cohesive whole. These Bridges help to flesh out the Bi-link Fabric, making it a living, breathing representation of your organization's data universe.

But the empowerment doesn't stop there. Your IT specialists, process managers, and other skilled employees can use the QBD API to automate the creation of Particles based on system events, process changes, or other significant occurrences. This means that your Bi-link Fabric continues to grow and evolve even as your business does.

This is the power of QBD. It's about more than just managing data; it's about giving everyone in your organization the tools and the agency to contribute to your data management efforts. It's about moving away from a centralized, top-down approach to data and towards a collaborative, bottom-up approach that leverages the unique insights and experiences of all your employees. In this way, QBD not only simplifies data management but also transforms it into a collective endeavor, driving innovation, collaboration, and growth.

Enhancing the Data Scientist's Role with QBD:

The image of a data scientist is often painted in broad, exciting strokes – an intellectual explorer charting the uncharted, using advanced algorithms and AI to unravel complex business problems. Yet, the reality for many data scientists is starkly different. Instead of diving deep into the realm of machine learning and AI, they find themselves knee-deep in the mundane task of data cleaning and preparation. Trying to make sense of disparate data systems, wrestling with inconsistent data types, and spending months, if not years, on understanding and normalizing data before they can even begin to apply their hard-earned expertise.

But what if we could shift that focus? What if, instead of spending a significant part of their time on data janitorial work, data scientists could dive straight into analysis, interpretation, and the application of machine learning models? That's where the Bi-link Fabric, built using Quantum Business Dynamics™ (QBD™), comes into play.

Imagine a scenario where the Bi-link Fabric acts as a unified, coherent, and easily accessible source of data for our data scientists. No longer would they need to spend months understanding why certain data is captured as integers in one system and as booleans in another. The Bi-link Fabric presents information about the agreements that cause the data to be created—that is, the real reasons 'why' action was taken, and this Bi-link data is always in a consistent, standardized format. This means that our data scientists can now start working with the information immediately, applying their machine learning models and AI algorithms to derive actionable insights.

And the Bi-link Fabric isn't just about making the data scientist's job easier - it's about making their work more impactful. With a unified, coherent view of the organization's agreements, our data scientists can uncover patterns and insights that would have remained hidden in a fragmented data landscape.

In essence, the Bi-link Fabric enables our data scientists to do what they do best - to extract value from information, to drive decision-making, and to contribute to the strategic goals of the organization. By redirecting their focus towards the Bi-link Fabric, we not only increase their job satisfaction but also elevate their role within the organization. They transition from being data janitors to true information explorers, delivering impactful insights and driving value.

Now, let's talk about the possibilities that unfold when data scientists are enabled to apply their expertise more effectively. Imagine a data scientist who, instead of wading through inconsistent and confusing data, can dive straight into analysis and problem-solving. Their expertise, once consumed by the overwhelming task of cleaning and understanding data, can now be channeled towards identifying trends, solving business problems, and driving innovation. This shift does more than just maximize the value of their analytical skills - it also increases their sense of engagement and fulfillment. They are no longer data janitors but vital contributors to the organization's strategic objectives.

The Bi-link Fabric also fosters a collaborative ecosystem that harnesses the collective knowledge of the organization. By providing an accessible and coherent view of the organization's agreements, QBD encourages everyone in the organization to contribute to its data management efforts. From every echelon of employees, everyone becomes a participant in this data ecosystem, adding their unique perspective and insights.

This collaborative approach helps to uncover new insights and opportunities for growth. When everyone in the organization can contribute to and learn from the Bi-link Fabric, it fosters a culture of continuous improvement and innovation. The organization becomes a living, learning entity, constantly adapting and evolving in response to new information and insights.

In essence, QBD and the Bi-link Fabric revolutionize the role of data scientists within the organization. They are no longer isolated figures wrestling with a tangled web of data. Instead, they become central figures in a collaborative ecosystem, leveraging their skills to drive innovation and growth. With QBD, we don't just transform our data management practices - we transform our organization's approach to knowledge, innovation, and growth.

Challenges and Considerations

Indeed, as with any transformative change, implementing QBD methodology is not without its challenges and considerations. However, these can be turned into opportunities with the right mindset, commitment, and best practices.

To begin with, ensuring organizational commitment to QBD methodology is crucial. Remember, QBD isn't just a new technology—it's a new way of thinking about data management. It requires a shift in mindset from top to bottom, from viewing data as a problem to be cleaned up to seeing it as a source of insights and strategic decisions. This commitment must be reinforced consistently and continuously, manifesting a culture that values data and its potential for driving growth and innovation. Key to this is communication—clearly articulating the benefits of QBD, celebrating successes, and addressing concerns openly and honestly.

Addressing potential barriers to adoption and integration is another critical consideration. These barriers might include technical challenges, resistance to change, or lack of understanding about QBD. The key is to anticipate these barriers and develop strategies to overcome them. For instance, technical barriers might be addressed through training and support, while resistance to change might be mitigated through engagement, education, and demonstrating the tangible benefits of QBD.

Finally, balancing QBD implementation with other data management best practices is essential. It's important to remember that QBD is not a replacement for these practices, but a complement to them. QBD doesn't require you to discard your existing data infrastructure—it works alongside it, enhancing its capabilities. This means you don't have to choose between QBD and other data management best practices—you can, and should, have both. In fact, QBD can help highlight areas where these practices can be improved or streamlined.

In essence, implementing QBD is a journey, not a destination. It requires ongoing commitment, adaptability, and a willingness to learn and grow. However, the rewards—in terms of enhanced data management, improved decision-making, and increased organizational agility—are well worth the effort.

Chain Blueprints: QBD Templates and Themes

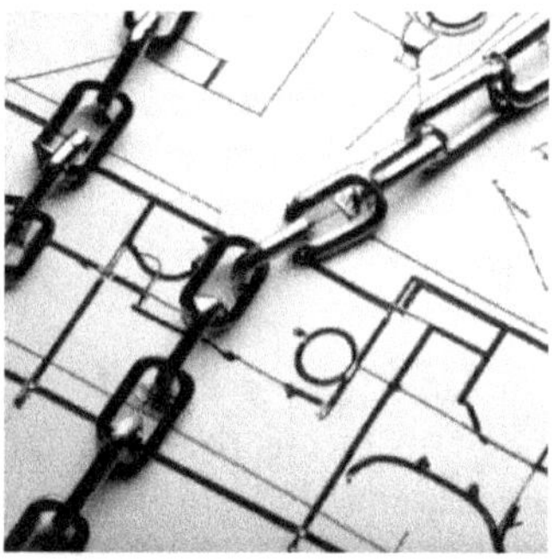

The adoption of Quantum Business Dynamics™ (QBD™) by businesses, while advantageous, requires careful planning and implementation. This is particularly true given the unique nature of each company's operations, necessitating a degree of customization to fully leverage the system's capabilities. However, considering the commonalities that exist across the 20 main NAIC industries, a strategy involving preconfigured templates and themes could significantly streamline the transition process, saving valuable time and resources.

When we look at industry-specific practices, procedures, and regulations, there are many shared elements across companies within each sector. For instance, a hospital in the Healthcare and Social Assistance sector and an auto repair shop in the Transportation and Warehousing sector might both find utility in an "insurance claim" theme. Despite operating in distinct industries, the processes they undertake when dealing with insurance claims could share similarities, thus making a common theme applicable.

Providing these preconfigured themes and templates, reflecting common Source, Target, and Bridge particles, can offer businesses a head start when adopting QBD. It grants them a foundational framework that they can then refine and expand upon to suit their specific operational needs.

However, the process of pre-configuration is not without its challenges. It demands an in-depth understanding of the nuances of each industry, ensuring that the preconfigured particles accurately align with the realities of businesses operating within those sectors.

Furthermore, while preconfigured themes and templates can expedite initial implementation, the inherent power of QBD lies in its adaptability and customization capabilities. To unlock its full potential, companies should give a little extra thought to training and system refinement, molding the QBD system to their unique operational requirements.

To conclude, the most efficient strategy for companies adopting QBD would likely involve a blend of preconfigured templates and themes, to kickstart their transition, followed by a rigorous customization process, ensuring the QBD system is perfectly tailored to their needs. This balanced approach could potentially harness the best of both worlds, accelerating adoption while preserving the adaptability that makes QBD so powerful.

Templates

Templates in Quantum Business Dynamics™ (QBD™) are essentially the blueprint for a company's Bi-link Fabric based on the common practices and regulations of a specific industry. They provide a starting point for companies adopting QBD by preconfiguring the fundamental particles (Sources, Targets, Bridges) necessary for the company's operations within its industry. Here are some key points to understand about templates:

1. **Industry-Specific Framework**: Templates are designed with industry-specific considerations in mind. This means they take into account the standard procedures, practices, and regulations shared across companies within a particular industry. For example, a template for the healthcare sector might preconfigure particles related to patient records, healthcare providers, insurance companies, and treatment procedures.
2. **Time and Resource Efficiency**: By providing a preconfigured setup, templates save businesses time and resources that would otherwise be spent on building the Bi-link Fabric from scratch. This

allows companies to begin operating within the QBD system more quickly and focus on customization and optimization based on their unique needs.

3. **Foundation for Customization**: While templates offer an initial setup, they are intended to be customized and built upon. Each company has its unique operational needs and the template serves as a starting point that can be modified and adapted to meet those requirements.
4. **Continuous Evolution**: Templates are not static; they are designed to evolve with the industry. As standard practices and regulations change, templates can be updated to reflect these changes, ensuring that companies remain compliant and up-to-date.
5. **Role of Platform Governor**: The platform Governor, such as Data Stewardship Solutions, Inc., plays a crucial role in developing and maintaining templates. They are responsible for understanding the trends and requirements of different industries and translating this knowledge into effective and efficient templates.
6. **Adoption Process**: When a company decides to implement QBD, one of the first steps is to select the appropriate industry-specific template. This template then serves as the foundation for the company's Bi-link Fabric, which can be customized and refined over time.
7. **Template Improvement**: The use of AI and machine learning by the platform Governor can analyze trends and performance across companies using a specific template. This analysis can lead to improvements and updates to the template, benefiting all companies within that industry.

In summary, templates play a crucial role in the adoption and implementation of QBD. They provide a balance between efficiency and customization, allowing companies to quickly adopt QBD while still maintaining the flexibility to adapt the system to their unique needs.

Themes

Themes in Quantum Business Dynamics (QBD™) represent preconfigured configurations for specific chains within the Bi-link Fabric. They offer a quick and efficient way to set up specific types of transactions or business operations that are common across different industries. Here are some key points to understand about themes:

1. **Standardized Configurations**: Themes provide a preconfigured

setup for specific types of transactions or operations. For instance, an "insurance claim" theme would preconfigure the necessary particles for a company to start a Bi-link chain for an insurance claim. This could include Source particles representing the claimant and the incident, Target particles representing the insurance payout, and Bridge particles representing the claim process.

2. **Cross-Industry Utility**: Unlike templates, which are industry-specific, themes can be applicable across multiple industries. For example, the "insurance claim" theme could be used by a hospital, an automobile body shop, or any other business that needs to process insurance claims.
3. **Efficiency and Consistency**: By providing a preconfigured setup, themes save businesses time and resources. They also ensure a level of consistency in how certain types of transactions or operations are conducted across different companies and industries.
4. **Foundation for Customization**: While themes offer a preconfigured setup, they are also intended to be customized and built upon to meet a company's unique needs. A company might start with a basic "insurance claim" theme, for instance, and then modify and expand it based on its specific claim processing procedures.
5. **Role of Platform Governor**: The platform Governor plays a crucial role in developing and maintaining themes. They are responsible for identifying common types of transactions or operations across different industries and translating this knowledge into effective and efficient themes.
6. **Theme Selection and Implementation**: When a company needs to set up a new type of transaction or operation within the QBD system, it can select an appropriate theme as a starting point. This theme then serves as the foundation for the company's Bi-link chain, which can be customized and refined as needed.
7. **Theme Improvement**: Just like templates, themes can also be improved over time. The platform Governor can use AI and machine learning to analyze trends and performance across companies using a specific theme and then make updates and improvements based on this analysis.

In summary, themes are an integral part of the QBD system. They offer a quick and efficient way to set up specific types of transactions or operations, while still maintaining the flexibility for companies to customize and adapt them based on their unique needs.

Typecasting Chains: The Drama of Strong vs. Weak

Quantum Business Dynamics™ (QBD™) introduces a significant concept in the realm of data management and analytics - Chain Typing. In a Bi-link chain, chain typing refers to the consistency of the types of Origins that bridge to Targets. This consistency can be 'strong' or 'weak', providing a level of flexibility and adaptability for data analysis. Chain typing opens up a new way of understanding and interpreting data, allowing businesses to extract more nuanced insights that are better aligned with their specific needs and goals.

Strongly Typed Chains

Strongly typed chains in QBD are those where the type of Origin that bridges to a Target is consistent throughout the chain. For instance, in a university setting, all faculty could consistently use either Certification or Confirmation as the type for students satisfying the requirements of each class. This consistency provides a level of standardization and predictability in the data, which can be particularly valuable for data analysis that requires precision and uniformity. In a strongly typed chain, data scientists can be confident in the consistency of the data, making it easier to identify patterns, trends, and correlations.

Weakly Typed Chains

On the other hand, weakly typed chains are those where the type of Origin that bridges to a Target can vary throughout the chain. Here, the flexibility of the chain allows for a broader range of data to be included, even if it doesn't strictly adhere to a single type. This flexibility can lead to the generation of diverse and potentially surprising insights, as data scientists can analyze a wider range of data points and their interconnections. Despite seeming less structured, weakly typed chains can reveal valuable patterns and correlations that would be missed in a more rigidly typed chain.

Leveraging Chain Typing for Business Benefits

The choice between strongly and weakly typed chains depends largely on the specific needs and goals of the business. Strongly typed chains might be preferable for businesses that value precision, consistency, and predictability in their data analysis. For example, a financial institution might opt for strongly typed chains to ensure the accuracy and reliability of its financial data.

In contrast, businesses seeking diverse insights and the potential for innovative discoveries might lean towards weakly typed chains. For instance, a marketing firm might find value in the patterns and trends emerging from a weakly typed chain of customer interactions.

Importantly, QBD allows businesses to choose between strongly and weakly typed chains, or even use a combination of both, depending on their specific needs. This flexibility enhances the utility of data and opens up new possibilities for data-driven decision-making.

So, chain typing in QBD provides a powerful tool for businesses to customize their data analysis, allowing them to extract more nuanced and targeted insights. Whether a business opts for strongly or weakly typed chains, or a combination of both, the power lies in the ability to tailor the approach to suit their unique needs and objectives. This adaptability is one of the many reasons why QBD is set to revolutionize the field of data management and analytics.

The Power of Target Transfers

We explained in the Target section how an Authority can transfer a Target to another party, who then becomes the new Authority. But now as you further understand the elegance of QBD, we can examine how this one little capability unlocks a most powerful Identity innovation.

Let's delve into a real-world scenario that encapsulates the potency of QBD, particularly the element of transferring Targets, in managing complex, interconnected situations.

Picture a patient, let's call them Alex, admitted to a hospital with a mysterious illness. The doctors are baffled by the symptoms, but eventually, one insightful doctor identifies the cause – exposure to a tainted pesticide. This revelation raises a grave concern: anyone who came into contact with this pesticide could fall ill as well.

Let's add another thread to this narrative. A farmer in a cannabis co-op is using QBD to manage the myriad regulatory checks from field to retail store. Now, let's connect these threads.

In the hospital's QBD system, Alex is known by a Source ID, let's say, Patient #12345. After the doctor's diagnosis, Alex reveals another identity, a different Source token – Loyal Customer #789 of a local hemp shop, where he suspects he might have come in contact with the tainted pesticide.

This is where the power of QBD comes into play. The Governor of the platform, upon receiving Alex's consent, creates a Target for Alex, representing him as a Carbon Based Life Form (CBLF). This Target is then bridged to the two Source tokens, effectively connecting Alex's identities as Patient #12345 and Loyal Customer #789.

What happens next is even more remarkable. Control of this Target is transferred to Alex, making him the new Authority. This enables researchers to trace back the Bi-link chain, right to the specific crop where a farmer was unwittingly sold the tainted pesticide, then forward to the retail stores and specifically to the customers who purchased products from that particular crop. This identification process can happen instantaneously via the pub/sub communication model inherent in QBD.

The benefits here are twofold. First, only the affected customers can be privately notified without the need to inform or involve the retail stores, distributors, or any other parties. This would be an impossibility with legacy data systems that require complicated court orders, inevitably exposing what could have remained private information. Second, should the initial contamination turn out to be a criminal act, the stealthy nature of an investigation properly leveraging the Bi-Link Fabric only by limited court order would be far less likely to alert the perpetrators, hence improving the chances of justice being served.

In the fascinating scenario of the hemp patient, a complex chain of events coalesced, demonstrating the unique capabilities of the QBD system and the power of Target transfer. Let's break down the situation and see how it reflects the rules in question.

i. Authority can transfer Target authority to another party: The patient was able to request the Governor create a Target for the patient, the Carbon Based Life Form (CBLF). With the patient's consent, this Authority was transferred to the patient himself, effectively making him the Authority of his own Target. This empowers the patient to control the information linked to his Target, granting him autonomy and control over his own data.

ii. CBLF Target Class created by the Governor: The Governor of the platform was responsible for creating the CBLF Target Class. This specific class allowed for the unique linking of the two Source tokens related to the patient, enabling the tracing of the patient's exposure journey.

iii. By request only by Owner: The creation of the Target and its subsequent transfer was initiated by the patient, the rightful owner of his own data. The patient's consent is a crucial element in this process, ensuring that personal data is handled respectfully and in compliance with privacy regulations.

iv. With transparency to CBLF: Throughout this process, transparency was maintained with the patient. He was made aware of the actions being taken and the implications of these actions, ensuring that he was informed and in control of his own data.

v. Must use consistent AA (Authorized Agency): To maintain the integrity of all related privacy rights, the related Bridges already have AA built in, which means that any time in the future, the patient can manage directly or via their chose AA their related Information Rights like opt-outs/ins.

vi. CBLF will have the capability to invalidate any Bridges to the CBLF Target: Upon becoming the Authority of his own Target, the patient gained the power to invalidate any Bridges to his Target. This means he can sever connections that he deems inappropriate or unwanted, thereby controlling access to his personal data.

This patient's journey through the QBD system vividly illustrates the potential of Target transfers and the powerful control it offers to the individuals involved. This capability can revolutionize data control, providing individuals with unprecedented autonomy over their personal data.

This scenario exemplifies the profound capabilities of QBD, not just in its operational aspects, but also in preserving privacy, enabling swift response, and potentially safeguarding public health.

QBD and DNA – Bridging Life Science and Data Science:

Quantum Business Dynamics™ (QBD™) offers a rich, layered landscape of data relationships and processes. Its intricate complexity is similar to the intertwining threads of a tapestry, where individual stories are woven together to create a beautiful, coherent whole. However, as we delve deeper into the fabric of QBD, we find that the tapestry analogy, while useful, is not entirely adequate. Let's enhance our understanding by exploring the more nuanced aspects of QBD - Chain Typing and Data Ownership.

Chain Typing: Beyond Linear Narratives

In our initial exploration of QBD, we understood Bi-link chains as linear sequences that follow a single, straightforward narrative. This perspective, while helpful for a beginner, is akin to looking at a single thread in a tapestry instead of stepping back to observe the larger picture from a distance, appreciating the detail without understanding the overall weave.

In reality, QBD chains are multidimensional, capable of forming multiple parallel narratives. Moreover, an Origin in one chain can bridge across to a Target in another chain, creating intersecting storylines that add depth and complexity to the data landscape.

Chain typing is the term we use to describe the nature of these narratives - whether they are strongly or weakly typed. As covered in detail earlier, a strongly typed chain or chain segment has Origins of the same Validation Type, creating a uniform and predictable narrative. On the other hand, a weakly typed chain features Origins of different Validation Types, leading to a more diverse and unpredictable storyline. For simplicity, we can think of an Origin's Validation Type as the Origin's type for now.

Vectorization: Unleashing the Power of Directionality

Just as Chain Typing and Data Ownership refine our understanding of QBD, so too does the concept of Vectorization. In the early stages of our exploration, we grappled with the basics of Binary Links and Chains, but QBD is far more nuanced, resembling an intricately knitted fabric rather than a simple line of stitches.

Vectorization introduces a layer of directionality and magnitude to the Binary Links and Chains, transforming the data landscape into a vibrant, multidimensional tapestry. It is the term we use to describe the orientation and strength of data relationships - whether they are strongly or weakly vectored.

A strongly vectored link or chain segment has a clear, unidirectional flow of data, creating a robust and predictable narrative. It's like a highway where traffic moves in one direction, allowing for efficient and speedy transmission of data.

On the other hand, a weakly vectored link exhibits a more diffused, multidirectional data flow, leading to a richer, albeit less predictable storyline. It's akin to a bustling city street where information moves in various directions, each interaction adding a new thread to the narrative.

In a broader context, Vectorization plays a pivotal role in shaping the strength and stability of QBD processes. It dictates how data is transferred, processed, and interpreted within the Quantum Business Dynamics™ framework, thus influencing the overall effectiveness and potential of QBD strategies. As we journey deeper into this realm, we continue to unveil the hidden dimensions of Vectorization and its profound implications for the evolving data landscape.

Vectorization, indeed, adds another layer of complexity to our understanding of Chain Typing, creating a more comprehensive picture of how data behaves within the QBD framework.

Vectorization, with its focus on the directionality and magnitude of data relationships, helps explain the dynamic nature of Chain Typing. A chain or chain segment, while being strongly or weakly typed, is not static but rather is subject to the influences of Vectorization.

In the case of a strongly typed chain, Vectorization can help identify sections that, despite having Origins of the same type, may have a diffused or multi-directional data flow - effectively behaving like a "weak" section. This occurs when the Targets within a strongly typed chain are vectorized, exhibiting multi-directional data links, thus introducing an element of unpredictability to an otherwise uniform narrative.

Similarly, a weakly typed chain, known for its diverse Origins and unpredictable storyline, can possess "strong" sections when vectorized Targets create a unidirectional flow of data, effectively streamlining the narrative within that section.

Thus, Vectorization allows us to view Chain Typing not merely as a binary classification of strong or weak but as a fluid, dynamic spectrum. It offers a more nuanced understanding of data behavior within chains, highlighting the intricate interplay of Chain Typing and Vectorization in shaping the data landscape within Quantum Business Dynamics™.

Data Ownership: Strength in Possessability

Another essential aspect we have covered is the strength of data ownership, determined by the possessability of a Binary Link. It's a concept that reflects the level of control or ownership an entity has over a particular Binary Link.

In QBD, the strength of ownership can be strong, where the owner of the Origin gets to possess the resulting Binary Link, adding to the strength of the chain. Conversely, the strength of ownership can be weak if the resulting Binary Link is possessed by an entity other than the Origin owner, which can weaken the chain.

In essence, we can think of QBD as a complex interplay between Chain Typing and Data Ownership. These intertwined concepts are crucial for understanding the intricacies of QBD and add depth to our understanding of this dynamic data landscape.

As we delve further into these concepts, we will discover how they shape the functionality, stability, and effectiveness of QBD processes, giving us even greater insight into the power and potential of Quantum Business Dynamics™.

Four Types of Strength:

These combinations will allow for 4 different possible Bi-link types. Let's explore each type:

SS Bi-links: Strongly Typed, Strong Ownership

An SS Bi-link is formed when an Origin from a strongly typed chain segment bridges to a Target, and the owner of the Origin can possess the resulting Binary Link. Both the typing of the chain and the ownership of the Binary Link are strong, hence the label "SS". This type of Bi-link offers high predictability and reliability due to its strong typing, combined with the strong ownership that ensures a firm control over the resulting Binary Link.

SW Bi-links: Strongly Typed, Weak Ownership

A Bi-link is classified as SW when the Origin comes from a strongly typed chain segment, but the resulting Binary Link, after bridging to the Target, is not owned by the Origin's owner. Although the chain segment's strong typing provides a level of consistency and predictability, the weak ownership introduces an element of flexibility and variability. The owner of the Origin does not have control over the resulting Binary Link, creating a unique dynamic that can lead to unexpected and potentially beneficial outcomes.

WS Bi-links: Weakly Typed, Strong Ownership

In a WS Bi-link, the Origin is derived from a weakly typed chain segment. Despite the Origin's weak typing, the owner still possesses the resulting Binary Link after bridging to the Target. This type of Bi-link is characterized by its robust ownership control, which compensates for the unpredictability and diversity inherent in the weakly typed chain segment. The resulting Binary Link, while somewhat unpredictable due to the weak typing of the chain segment, remains under the control of the Origin's owner.

WW Bi-links: Weakly Typed, Weak Ownership

A WW Bi-link presents the most flexible and dynamic scenario. The Origin comes from a weakly typed chain segment, and the resulting Binary Link, after bridging to the Target, is not owned by the Origin's owner. This type of Bi-link is characterized by a high level of variability and potential due to both the weak typing and weak ownership. This can lead to a wide range of outcomes, as the resulting Binary Link's behavior is influenced by a broad set of factors.

Therefore, understanding these four types of Bi-links – SS, SW, WS, WW – allows for a more nuanced grasp of QBD's potential. Each type provides a different set of characteristics that can be leveraged to create a variety of data narratives and achieve diverse objectives.

Understanding the distinctions between SS, SW, WS, and WW Bi-links isn't a matter of intellectual curiosity alone. Instead, it's an essential piece of the puzzle in achieving a robust and versatile understanding of Quantum Business Dynamics™ (QBD™). This knowledge greatly enhances the depth and breadth of insights one can derive from Bi-link chains, leading to more accurate and actionable outcomes.

Let's consider the example of data analysis within a business context. As a data analyst, if you're familiar with these types of Bi-links, you'll have a better understanding of the data's structure and behavior. This enables you to make more informed predictions and generate more relevant insights that can guide business decisions.

For instance, if you encounter an SW Bi-link, you'll know that the Origin comes from a strongly typed chain segment but the resulting Binary Link isn't owned by the Origin's owner. This knowledge can inform your expectation about the predictability of the data's behavior (due to the strong typing) and the potential variability of the data ownership, providing a more nuanced context for your analysis.

Moreover, the ability to recognize and understand these Bi-link types could guide the design and implementation of future data structures within your organization. It could help identify which structures are more likely to yield desired outcomes, or which structures might offer the most flexibility or control.

Furthermore, this understanding can also enhance collaborative efforts. When working with other teams or organizations that employ QBD, understanding these Bi-link types can facilitate more effective communication and alignment, ensuring all parties have a shared understanding of the data's underlying structure and behavior.

That is why understanding these four types of Bi-links is far from being trivial. It's a fundamental aspect of QBD that can greatly enrich the quality of your data analysis and the subsequent insights derived from it. It empowers you to not only understand the current data narratives better but also to predict future narratives and adapt strategies accordingly.

Probabilistic vs. Deterministic

Probabilistic and deterministic are two contrasting concepts that come up often in machine learning and data science. Understanding the differences between these terms is crucial in choosing the right predictive model for your data.

A deterministic model is one where the outcomes are precisely determined through known relationships among states and events, without any room for randomness. In other words, given a specific input, the output will always be the same. A common example is a simple arithmetic operation like addition. If you add 2 and 3, the result will always be 5, without any uncertainty.

Probabilistic modeling, on the other hand, embraces the uncertainty inherent in predicting outcomes. Instead of providing a single, definitive outcome, probabilistic models yield a range of possible outcomes, each associated with a specific probability. This reflects the inherent randomness or uncertainty in the data or the event being predicted.

Now, let's apply these concepts to Quantum Business Dynamics™ (QBD™) and the Bi-link types (SS, SW, WS, and WW). Understanding the characteristics of these links provides a structured context for feature selection, a crucial step in the machine learning process.

Each Bi-link type represents a unique combination of chain typing and data ownership, which could be considered as features in a machine learning model. For example, an SS link, being a strongly typed link with data ownership belonging to the origin's owner, might be associated with more consistent, predictable behavior in the data, which aligns more with a deterministic perspective. On the other hand, a WW link, being a weakly typed link with data ownership not belonging to the origin's owner, might show greater variability, aligning more with a probabilistic perspective.

Understanding these Bi-link types could also contribute to the development of more sophisticated probabilistic models. By incorporating the structure and behavior of these link types into the design of these models, we could potentially enhance their ability to capture the nuances in the data and therefore increase their predictive accuracy.

For instance, a Bayesian network, a type of probabilistic graphical model, could be designed with nodes representing the different Bi-link types, and the edges could represent the probabilistic relationships between them. This could provide a powerful way of capturing and representing the complex dependencies in QBD data, leading to more nuanced and accurate predictions.

To sum up that thought, comprehending the Bi-link types in QBD can significantly enhance the process of feature selection and probabilistic modeling in machine learning. Understanding the differences between probabilistic and deterministic models is essential in this process, as it guides the selection of the appropriate modeling strategy to yield the most accurate and insightful predictions.

Now…we once again find ourselves on the precipice of an intriguing convergence with Natural Law and we are going to learn a great deal about your business in ways you never before dreamed. As we've seen with the parallels between chemistry charges and ledger charges, QBD's theoretical framework often reflects fundamental principles seen in the natural world. A new and striking example of this correlation emerges when we compare the predictive modeling of DNA particles in biology to the predictive modeling of Bi-link Particles in QBD. The similarities between these two concepts not only enrich our understanding of QBD, but also open up new possibilities for predictive modeling in business, a territory previously uncharted.

The essence of DNA's predictive power lies in the four nucleobases: Adenine (A), Thymine (T), Guanine (G), and Cytosine (C). These four bases pair up in a specific manner: A with T, and G with C, forming the building blocks of all genetic information. The stability of these pairings, governed by the number and type of hydrogen bonds between them, has been a key factor in predicting the structure and functionality of genes.

Similarly, in QBD, we classify Bi-links into four types based on two parameters: the strength of the typing of the chain, and the ownership of the resulting Binary Link. These types are: SS (Strong-Strong), SW (Strong-Weak), WS (Weak-Strong), and WW (Weak-Weak).

Comparing the distinctions between these four basic and parallel elements of both DNA and QBD can result in some fascinating additional considerations for predictive analysis of models. There are three strong comparisons that can be made here:

Chain Typing (Strongly typed vs. Weakly typed chains): This difference in QBD could be likened to the structural differences between purines (adenine and guanine) and pyrimidines (cytosine and thymine) in DNA. In both cases, the differing configurations lead to different properties and behaviors. For example, just as purines and pyrimidines pair differently due to their structures, strongly typed and weakly typed chains in QBD may interact differently with other elements.

Vectorization: This process in QBD, which allows for different resulting Binary Link types based on the Origin type, could be analogous to the base pairing rules in DNA. Just as the specific pairing of nucleobases (A-T and G-C) determines the genetic code, the vectorization rules in QBD govern the transformation of Origins to Targets, shaping the ultimate structure and function of the system.

Ownership Possession (Owner vs. Non-owner of the Origin): This distinction in QBD could correspond to the number of hydrogen bonds formed between nucleobases in DNA. Just as adenine (A) always pairs with thymine (T) via two hydrogen bonds, and guanine (G) always pairs with cytosine (C) via three hydrogen bonds, the rules of ownership possession in QBD determine the strength of the connection between the Origin and the resulting Binary Link.

The comparison between DNA base pairs and QBD Bi-link types is particularly striking when we consider them as counterparts with similar functionality:

- SS (Strong-Strong): This could be likened to the G-C pair in DNA. The G-C pair, composed of two purine bases, is the most stable due to the three hydrogen bonds it forms. Similarly, SS represents the strongest type of Bi-link, with both strong typing of the chain and strong ownership of the resulting Binary Link.
- SW (Strong-Weak): This could be analogous to the A-G pair. The larger purine structure (akin to the strong typing of the chain in SW) gives it stability, but the fewer number of hydrogen bonds (akin to the weak ownership possession in WS) makes it less stable than G-C (SS), but more stable than T-C (SW).
- WS (Weak-Strong): This could correspond to the T-C pair. The smaller pyrimidine structure (akin to the weak typing of the chain in WS) reduces its stability, but the additional hydrogen bond (akin to the strong ownership possession in WS) adds some stability, making it less stable than A-G (SW) but more stable than A-T (WW).
- WW (Weak-Weak): This can be likened to the A-T pair in DNA. A-T pair, composed of two pyrimidine bases, is the least stable due to the two hydrogen bonds it forms. Similarly, WW represents the weakest type of Bi-link, with both weak typing of the chain and weak ownership of the resulting Binary Link.

These comparisons provide us with a novel way of understanding the stability and dynamics of Bi-links in QBD systems, drawing upon the well-established principles of DNA base pairing. This framework can guide data scientists and QBD practitioners in developing more accurate predictive models and strategies.

The predictive power of biology, particularly in understanding DNA, can now be applied to business data using these models. For instance, just as the stability of G-C and A-G pairs in DNA could predict the structure of a gene, the stability of SS and SW Bi-links can be used for deterministic predictions for the potential outcomes in a QBD system, and the other two, WS and WW for more probabilistic.

To give a concrete example, consider a scenario where a data scientist is trying to predict customer behavior in a retail environment. If the data scientist observes a pattern of SW Bi-links (akin to A-G pairs), they might predict a relatively stable but somewhat variable customer behavior. However, if they observe a pattern of WS Bi-links (akin to T-C pairs), they might predict a more volatile customer behavior due to the lower stability.

By borrowing predictive principles from biology and applying them to QBD, data scientists can enhance their predictive modeling techniques, and businesses can gain a deeper understanding of their operations, leading to more informed decision-making. This is yet another testament to the profound interplay between Natural Law and QBD, and a promising sign of the innovative applications that lie ahead.

Unleashing the Power of QBD: Cross-Disciplinary Applications and Insights

Quantum Business Dynamics™ (QBD™) is a paradigm shifting methodology, capable of breaking down and predicting behaviors in business data in a way few systems can. The strength of QBD lies not just in its innovation, but also in its reflection of principles that underpin a wide array of scientific disciplines, from biology and sociology to psychology and beyond.

The beauty of this multidisciplinary overlap is not merely in its intellectual richness. It opens the door to the cross-pollination of machine learning models between disciplines, a prospect that is as exciting as it is groundbreaking.

Bridging the Gap: Machine Learning Models Across Disciplines

Consider this: a machine learning model developed for understanding the genetic dynamics of mutation within a species, under the right circumstances, could be applied directly to QBD's Bi-link Fabric. This is possible given the correct mapping from the model's required data feature sets (like DNA base pairs A-C-G-T) to QBD's combinations of links (SS, SW, WS, WW). Such an application could shed light on business "mutations," such as a change in product lines or organizational structures, and their potential impacts on a company's survival and success.

This cross-pollination extends beyond biology. The field of psychology, for instance, boasts a wealth of models that analyze patterns of individual and group behaviors. These models, when applied to QBD data, could provide fresh insights into customer behavior, team dynamics, or leadership styles.

Similarly, sociological models that study societal trends and movements could offer a macro perspective on market shifts and industry trends when mapped onto a QBD Bi-link Fabric.

Even fields as diverse as environmental sciences could offer valuable models, such as those predicting climate change impacts or ecosystem dynamics, which, when applied to business data, could forecast market volatility or anticipate the ripple effects of changes within an industry ecosystem.

The Value Proposition: QBD's Dollar Impact

The real question is, how does this translate into monetary value for businesses? We're not just talking about incremental improvements here. With the proper application of these interdisciplinary machine learning models, businesses could potentially unlock an estimated $100 billion in value across various industries.

This estimate might seem staggering, but let's consider this: businesses spend billions each year trying to predict market trends, understand customer behavior, and adapt to industry changes. With the ability to leverage existing, proven models from diverse scientific disciplines, the speed, accuracy, and depth of these insights could dramatically increase, leading to better decision-making and more efficient resource allocation.

The Path Forward: Seizing the Opportunity

The potential for cross-disciplinary applications of QBD is immense. The treasure trove of existing machine learning models across various scientific fields is waiting to be harnessed. All it takes is the right mappings, an understanding of the principles that translate across fields, and a willingness to venture into uncharted territory.

In this unexplored landscape, the opportunities are boundless. Businesses could gain unprecedented insights into their operations, customers, and markets. Scientists could find new applications for their models and uncover surprising parallels between their fields and the world of business. And together, we could advance our understanding of complex systems in ways we've yet to imagine.

It's not an overstatement to say that QBD could revolutionize our approach to business data analysis, not just within the confines of business, but in a way that intertwines with the broader tapestry of scientific understanding. This is the power and potential of Quantum Business Dynamics™.

The Quantum Bridge: A Question of Refraction, Not Equivalence

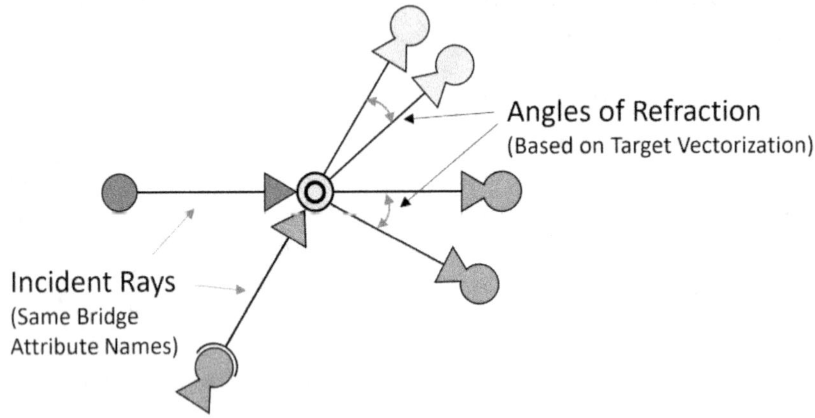

In the domain of Quantum Business Dynamics™ (QBD™), our understanding of the relationships and interactions between Particles is key to unlocking the full potential of the system. One of the foundational aspects of this understanding lies in the equation:

$$A => B = C$$

...or Origin (A) is bridged (=>) to a Target (B) resulting (=) in a Binary Link (C). However, this equation has been met with some misunderstandings and misinterpretations, particularly regarding the nature of C, the resultant Binary Link.

The equation represents a simple but profound concept: Origin A, through Bridge => to Target B, results in Binary Link C. However, some have suggested that C itself is the Bridge, or that it represents the Bridge. This interpretation, although seemingly logical at first glance, significantly undermines the depth, dynamism, and versatility of the QBD process.

In JavaScript, a language familiar to many, the difference between the equal sign (=) and the double equal sign (==) is profound. The former indicates a yield, a new value resulting from a combination of elements, which must consider a certain property or properties of both Particles being bridged. The latter suggests a Boolean equivalence, a true or false condition. It's clear that in the context of QBD, A combined with B does not equate to C – it yields C. The subtle but significant difference lies in understanding C not as the Bridge, but as the outcome of the agreement that is the interaction between A and B.

Why is this distinction so vital? Let's delve into the dynamics of the equation. In QBD, the => symbol represents an agreement between two parties, the Bridge. Binary Link C encapsulates the outcomes of this agreement. However, viewing C as the Bridge or agreement itself is a limited perspective.

The true beauty and complexity of QBD lies in the fact that Origin A is not a constant. It brings with it a vector, a directionality which influences how the agreement passes through Target B. Binary Link C's attributes can vary dramatically depending on this vectorization, as we see in the previous section. Conversely, all Bridges (=>) between A and B will share the same set of attributes, while the resulting Binary Links (Cs) will likely not. This distinction is crucial. If C were the Bridge, it would restrict the depth and flexibility of the system.

To better understand this, consider the physical concept of refraction. When light passes from one medium to another, its direction changes - it refracts. This shift in direction is influenced by the angle of the incoming light (incidence) and the properties of the medium it's passing through.

In the realm of QBD, Origin A is akin to the angle of incidence - it carries a specific vector or directionality. Target B is like the medium - its properties influence the direction of A. The resultant Binary Link C is similar to the refracted light beam - its attributes are the result of the interplay between A and B.

Just as light takes different paths depending on its angle of incidence and the medium it's passing through, different A's passing through B by way of the exact same type of Bridge can lead to different Cs. The refraction of light is not merely the medium it passes through, but the outcome of its interaction with the medium.

Viewing C as the Bridge would be akin to saying that the refracted light is the medium it passed through - an oversimplification that overlooks the interplay of factors that determine its final direction.

This analogy illuminates why C cannot be considered as the Bridge. It is the result of the interplay between A and B, influenced by their unique properties and interactions. Understanding this difference is crucial to fully appreciate the elegance and power of Quantum Business Dynamics™.

By embracing the dynamic interplay between Origins, Bridges, Targets, and Binary Links, we can move beyond simplifications and unlock the full potential of QBD. Only then can we truly harness its capabilities to yield insightful and transformative outcomes in the realm of business data.

The Quantum Bridge and Durability: Shaping the Lifespan of Light

Quantum Business Dynamics™ (QBD™) introduces Bridge Durability, a property that establishes the lifespan of a Bridge. It's a feature that can be set at the Target level, determining Durability for all subsequent Bridges, or it can be defined individually for each Bridge. This unique feature of QBD has vast implications across various domains.

Have you ever wondered about Schrödinger's cat? The mind-bending thought experiment postulates that a cat in a box could be simultaneously alive and dead, a state known as superposition in quantum mechanics. Bridge Durability in QBD similarly sets the stage for a superposition-like state. A Bridge could be either on the verge of expiration or Timeless, triggering a cascade of potential realities for any given agreement, each unfolding in a different manner. This uncertainty is a playground for AI and ML systems to consider diverse scenarios and derive fresh insights.

To understand this concept, let's examine the case of Republic Pictures, the studio that unknowingly allowed the rights to the classic film "It's a Wonderful Life" to lapse. Having originally set the rights to be Temporal, they missed the opportunity to renew and consequently lost a significant return on investment when the movie gained renewed popularity decades later. With QBD, a Durability feature for such rights could ensure that important renewals never fall through the cracks, safeguarding the potential for future profits.

On the other end of the spectrum, consider the case of a widow falling on hard times. A complex estate, potential pensions, old bonds — all these assets might possess lasting value (Timeless Durability), yet often go unnoticed, leaving those who need them most unsupported. Today, there is a staggering $58.5 billion of unclaimed money in the United States alone. With a robust Bridge infrastructure set to Timeless, the discovery of such assets becomes seamless, potentially providing much-needed relief to individuals like our hypothetical widow, as well as reducing societal burden.

Durability, in the world of QBD, is the regulator of light — it determines how long the light can shine through the prism of a Target. From preventing financial missteps like those of Republic Pictures to discovering timeless assets hidden in the fabric of our lives, Durability illuminates diverse opportunities. As we delve deeper into the wonders of QBD, let's explore how far this light can travel.

Saving the Best for Last: SNC-Activation

Quantum Binary Dynamics™ (QBD™) holds a myriad of possibilities, bringing transformative features that are not just novel but can truly revolutionize multiple industries. One of these standout features is Social Networking Communication (SNC) activation, a feature that adds an unprecedented layer of transparency, engagement, and excitement. Here, we illuminate the brilliance of this feature, tracing the path of the "light" as it passes through the prism of a QBD Target.

Adding a new data use to the Consent Name Service (CNS) is a straightforward process. Let's take, for instance, the recent introduction of a new data use for "Social Networking Communication (SNC)". Businesses could make a query like, "Can I use the data for this Binary Link token in these specified jurisdictions for SNC?" In an instant, the system provides a clear answer—a '0' indicating you cannot do that, or a '1' indicating you can.

Now, imagine coupling this functionality with a publish/subscribe (pub/sub) service that can trigger messages based on subscribed events. This service enables monitoring of Binary Link token creation in relation to a specific Target or group of Targets. When the Authority creates a Target and chooses to set the 'makeSocial' attribute to true, all previously confidential information found in the resulting Binary Links from that Target are earmarked for potential social networking communication purposes.

An actor, such as Data Stewardship Solutions (DSS), with the appropriate authorization, and with the Rights Resource Management Agent in place from the associated Bridge, can subscribe to all of these events. The real magic begins when DSS hosts a message queue that, upon invitation, subscribes on a token-by-token basis for any Binary Links created from SNC-activated Targets. This process results in an aggregation of data which, upon gaining necessary permissions from the CNS, can be published to a stream of channels like RSS for SNC.

This allows for a thrilling new way to engage fans and followers. Imagine being able to subscribe to the progress of a vineyard's seasonal journey, tracking data from vine to wine. The predictions on quality based on all sorts of bi-link fabric insights such as growing condition, heritage, and harvesting care could drive up anticipation for specific batches!

Or maybe you're interested in outer space projects, legislation in development, pre-season sports training, or the launch of a new craft beer.

With SNC-activation, these processes are no longer obscured behind industry walls. The light of information can shine farther than ever before, reaching passionate followers and allowing businesses to share their journey in a way that builds anticipation and engagement. SNC transforms industry processes into social narratives, facilitating a powerful and dynamic conversation between businesses and their audience.

This exciting feature also opens up possibilities for cottage industries and influencers. Aficionados can subscribe to their favorite projects, start their own YouTube channels, and build their brand around their passion, buoyed by the streams of information provided by QBD's SNC-activated Targets.

In conclusion, SNC-activation is like a master switch that opens up industries to the social sphere, adding depth to data and transforming industry professionals into potential celebrities. Through QBD, the flow of information becomes a brilliant stream of light, refracted by the prism of Targets, spreading far and wide, illuminating the world with a transparency, engagement, and excitement that was hitherto unthinkable. We truly saved the best for last!

SECTION II – BRIDGING BETWEEN

Bridging Between Groups

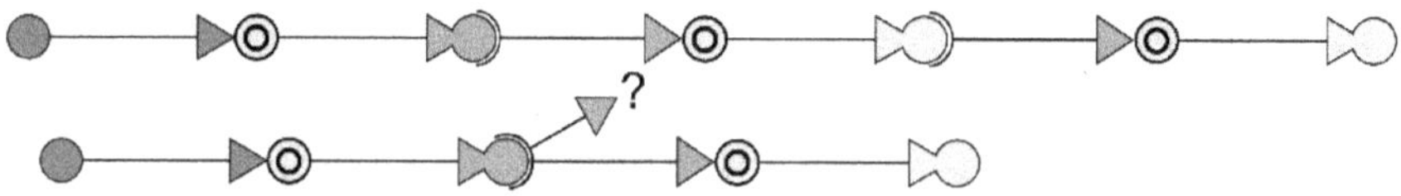

Change Management

Change management is a structured approach to transitioning individuals, teams, and organizations from a current state to a desired future state and it's a perfect way to manage your organization's adoption of Quantum Business Dynamics™.

The following steps can be applied to effectively implement a new business methodology between internal groups of a company.

1. Create a compelling case for change: The rationale for the new business methodology must be clearly communicated and understood by all stakeholders.

   I. Introduction

Explaining how the Quantum Business Dynamics™ (QBD™) methodology can help your organization achieve its goals and objectives is always a good way to start when building on your adoption. To accomplish this, start by communicating the need for change, e.g., to keep up with industry changes, improve financial performance, or increase efficiency. Then, explain how QBD can help address these challenges and provide a clear vision of how the organization will look with QBD.

To make a compelling case for change using popular business styles, consider using the following techniques:

- Tell a story: Use a narrative to explain why the current approach is no longer effective and how QBD can provide a better way forward. Use examples to illustrate the benefits of QBD and show how it has helped other organizations in similar situations.
- Use data: Use data to demonstrate the impact of QBD on key metrics such as revenue, costs, and customer satisfaction. Use visual aids such as graphs and charts to make the data more accessible and understandable to stakeholders.
- Use emotional appeals: Use emotional appeals to connect with stakeholders and create a sense of urgency around the need for change. For example, highlight the risks of not adopting QBD or show how QBD can help employees feel more engaged and fulfilled in their work.

II. Identify the need for change

Explain the current challenges or issues that the organization is facing and how QBD can help address these challenges. To do this, conduct a thorough analysis of the current state of the organization, including its strengths and weaknesses. Then, show how QBD can help the organization overcome its weaknesses and take advantage of its strengths. Use specific examples to illustrate the benefits of QBD.

To identify the need for change using popular business styles, consider using the following techniques:

- Conduct a SWOT analysis: Use a SWOT analysis to identify the strengths, weaknesses, opportunities, and threats facing the organization. Use the results of the analysis to show how QBD can help the organization overcome its weaknesses and take advantage of its strengths.
- Use case studies: Use case studies to illustrate how QBD has helped other organizations in similar situations. Use the case studies to show how QBD can help the organization achieve its goals and objectives.
- Conduct a stakeholder analysis: Use a stakeholder analysis to identify the key stakeholders who will be impacted by the change initiative. Use the results of the analysis to tailor the message to each stakeholder group and address their specific concerns.

III. Define the new business methodology

Explain what QBD is and how it differs from the current approach. To do this, provide a clear and concise explanation of the key principles of QBD, such as its focus on systems thinking, quantum leadership, and stakeholder engagement. Use simple language and provide examples to help stakeholders understand how QBD works and why it is better than the current approach.

To define the new business methodology using popular business styles, consider using the following techniques:

- Use metaphors: Use metaphors to explain complex concepts in a simple and memorable way. For example, compare QBD to a symphony orchestra where all the parts work together to create a beautiful sound.
- Use analogies: Use analogies to draw comparisons between QBD and something that stakeholders are familiar with. For example, compare QBD to a GPS system that provides directions to help the organization reach its goals.
- Use storytelling: Use storytelling to illustrate the key principles of QBD and show how they can be applied in real-world situations. Use stories to create a sense of empathy and connection with stakeholders.

IV. Identify the benefits

Discuss the benefits of QBD for the organization and its stakeholders, including increased efficiency, better decision-making, and improved financial performance. To do this, quantify the benefits wherever possible, e.g., by showing how QBD can increase revenue, reduce costs, or improve customer satisfaction. Also, be sure to address any potential drawbacks of QBD and explain how they can be mitigated.

To identify the benefits of QBD using popular business styles, consider using the following techniques:

- Use testimonials: Use testimonials from other organizations or stakeholders to illustrate the benefits of QBD. Use the testimonials to show how QBD has helped other organizations achieve their goals and objectives.
- Use metrics: Use metrics to quantify the benefits of QBD, such as increased revenue, reduced costs, or improved customer satisfaction. Use visual aids such as graphs and charts to make the metrics more accessible and understandable to stakeholders.

- Use ROI analysis: Use a return on investment (ROI) analysis to calculate the financial benefits of QBD. Use the analysis to show stakeholders how QBD can generate a positive return on investment and provide a clear picture of the financial benefits of the change initiative.

V. Address potential concerns and objections

Anticipate potential concerns or objections from stakeholders, such as perceived risks or lack of understanding of the new methodology. To do this, conduct a stakeholder analysis to identify potential concerns and objections. Then, develop strategies to address these concerns or objections, such as providing additional training, conducting workshops, or addressing concerns one-on-one with stakeholders.

To address potential concerns and objections using popular business styles, consider using the following techniques:

- Use case studies: Use case studies to show how QBD has helped other organizations overcome similar concerns or objections. Use the case studies to demonstrate how QBD can be successfully implemented and how any risks or concerns can be mitigated.
- Use social proof: Use social proof to show that QBD is a widely accepted and successful methodology. Use references to reputable sources, such as industry leaders or well-respected experts, to show that QBD is a proven methodology.
- Use persuasion techniques: Use persuasion techniques, such as framing the message in a positive light or using vivid language, to overcome objections or concerns. Use the techniques to create a sense of urgency and commitment among stakeholders.

VI. Develop a communication plan

Identify the key stakeholders who need to be informed about the change initiative, such as executives, managers, and employees. Then, develop a communication plan that outlines what information needs to be shared, when it needs to be shared, and how it will be shared. Use multiple communication channels, such as town hall meetings, email updates, and training sessions, to ensure that stakeholders receive the information they need in a timely manner.

To develop a communication plan using popular business styles, consider using the following techniques:

- Use a communication matrix: Use a communication matrix to identify the key stakeholders, the messages that need to be communicated to each stakeholder group, and the communication channels that will be used. Use the matrix to ensure that all stakeholders receive the information they need in a timely manner.
- Use a communication hierarchy: Use a communication hierarchy to ensure that messages are communicated in a clear and concise manner. Use a top-down approach, where executives communicate the key messages to managers, who then communicate the messages to employees.
- Use a feedback loop: Use a feedback loop to ensure that stakeholders are engaged and receptive to the change initiative. Use feedback mechanisms, such as surveys or focus groups, to gather feedback and make adjustments to the communication plan as needed.

VII. Conclusion

Summarize the key points of the compelling case for change by emphasizing how QBD will help the organization achieve its goals and objectives. Encourage stakeholders to support the change initiative and become advocates for the new business methodology by highlighting the benefits and providing ongoing support and training. Finally, establish a plan for monitoring and evaluating the implementation of QBD to ensure that it is successful and that stakeholders continue to be engaged and supportive.

To conclude a compelling case for change using popular business styles, consider using the following techniques:

- Use a call to action: Use a call to action to encourage stakeholders to support the change initiative. Use language that is persuasive and inspiring, such as "Join us in creating a brighter future for our organization with QBD."
- Use storytelling: Use storytelling to illustrate how QBD can help the organization achieve its goals and objectives. Use examples to show how QBD has helped other organizations in similar situations and create a sense of hope and optimism among stakeholders.
- Use a plan for monitoring and evaluation: Use a plan for monitoring and evaluation to ensure that QBD is successful and that stakeholders continue to be engaged and supportive. Use metrics and data to measure the success of the change initiative and make adjustments as needed to ensure that the organization continues to move in the right direction.

2. Develop a change management plan: This plan should identify the goals, objectives, and strategies for implementing the new business methodology.

I. Introduction

To introduce the need for change within the organization, start by explaining the current challenges faced by the organization that QBD can help solve. For example, if the company is struggling with meeting customer demands due to an inefficient process, highlight this as the need for change. Then, clearly explain how QBD can help address this challenge, using a relatable example or case study to illustrate the benefits.

For instance, "At ABC Company, we've noticed a pattern of customer complaints about delayed delivery times due to an inefficient production process. Quantum Business Dynamics™ can help us streamline this process, reduce lead times, and deliver products to our customers faster. For example, XYZ company has successfully implemented QBD and reduced their lead times by 40%, resulting in a 20% increase in customer satisfaction."

II. Define the goals and objectives

To define the goals and objectives of the change initiative, involve key stakeholders and conduct a thorough analysis of the current state of the organization. Use a SWOT analysis to identify the organization's strengths, weaknesses, opportunities, and threats. Based on this analysis, identify specific areas for improvement and establish SMART goals and objectives that will help the organization achieve its desired outcomes.

For example, "Our SMART goal is to reduce lead times by 50% within the next 6 months. Our objectives include implementing QBD methodology, optimizing our production process, and training employees on the new methodology. These goals and objectives will help us improve efficiency, enhance productivity, and reduce costs."

III. Identify the stakeholders

To identify the stakeholders who will be impacted by the change initiative, conduct a stakeholder analysis and gather input from employees, managers, and executives. Use a stakeholder mapping tool to identify key stakeholders and their level of influence and interest in the initiative. Develop strategies to address their concerns and needs, such as providing training and resources to help employees adjust to the new methodology.

For example, "Our stakeholder analysis revealed that our employees are concerned about the new methodology being too complex and time-consuming. To address this concern, we will provide hands-on training and coaching to ensure that they are comfortable with the new methodology."

IV. Develop the implementation strategy

To develop the implementation strategy, identify the appropriate QBD tools and techniques that align with the organization's goals and objectives. Use the "PDCA" cycle to develop a plan for implementing QBD, test it out, evaluate the results, and make adjustments as necessary. Develop a training plan that includes hands-on training and coaching to ensure that employees are comfortable with the new methodology.

For example, "Our implementation strategy includes using process mapping and statistical process control tools to optimize our production process. We will also provide hands-on training and coaching to ensure that our employees are comfortable with the new methodology. As we implement these changes, we will regularly evaluate the results and make adjustments as necessary."

V. Develop a communication plan

To develop a communication plan, use the "5W1H" framework to answer the questions of who, what, when, where, why, and how communication will take place. Develop a communication plan that includes regular updates, feedback sessions, and opportunities for stakeholders to ask questions and provide input.

For example, "Our communication plan includes weekly email updates to all stakeholders, monthly town hall meetings to provide updates and gather feedback, and feedback sessions after each training session to address any concerns. We will also provide a QBD handbook to all employees, outlining the new methodology, and its benefits."

VI. Develop a monitoring and evaluation plan

To develop a monitoring and evaluation plan, use Key Performance Indicators (KPIs) to track progress towards achieving the goals and objectives identified in step II. Establish regular evaluation checkpoints to analyze the data and make adjustments as needed. For example, if the goal is to reduce the time it takes to complete a certain process by 50%, track the time it takes before and after implementation and adjust the methodology as necessary.

For example, "Our KPIs include the average lead time, customer satisfaction ratings, and employee feedback. We will track these metrics before and after implementation to ensure that we are achieving our goals and objectives. We will hold regular evaluation checkpoints to review the data and make adjustments as needed."

VII. Conclusion

To conclude the change management plan, summarize the key points of the plan and emphasize the benefits of QBD. Encourage stakeholders to support the change initiative and become advocates for QBD by highlighting the benefits and providing ongoing support and training. Finally, establish a plan for monitoring and evaluating the implementation of QBD to ensure that it is successful and that stakeholders continue to be engaged and supportive.

For example, "In conclusion, QBD can help us achieve our goals of improving efficiency, enhancing productivity, and reducing costs. By defining SMART goals and objectives, identifying stakeholders, developing an implementation strategy, creating a communication plan, and monitoring and evaluating the implementation, we are confident that we can successfully implement QBD and achieve our desired outcomes. We encourage all stakeholders to support this change initiative and become advocates for QBD."

3. Engage stakeholders: Stakeholder engagement is critical for the success of the change initiative. Involve all relevant internal groups in the planning and implementation process.

Imagine you're leading a team tasked with implementing Quantum Business Dynamics™ (QBD™) between groups within the same company. You're excited about the potential for QBD to improve efficiency, enhance productivity, and reduce costs, but you know that stakeholder engagement is critical to realizing these benefits. Here's how you can engage stakeholders effectively:

I. Introduction

The first step in engaging stakeholders is to introduce the purpose of the change initiative and the importance of stakeholder engagement in achieving success. You can use a framework like the "what-why-how" model to explain what QBD is and why it's important, then outline how stakeholder engagement is critical to realizing the benefits.

For example, you might say something like, "At the heart of any successful change initiative is stakeholder engagement. Without buy-in and involvement from all relevant internal groups, the implementation of QBD could fall short of its potential. That's why it's critical to engage stakeholders at every stage of the process."

II. Identify stakeholders

The next step is to identify the stakeholders who will be impacted by the change initiative and their level of readiness for change. This could include employees, managers, executives, and other relevant internal groups. You can use a stakeholder mapping tool to assess each stakeholder's level of influence and interest in the initiative.

For example, you might say something like, "Stakeholder mapping is a powerful tool that can help you identify who your stakeholders are, and how they will be impacted by the change initiative. By understanding each stakeholder's level of influence and interest, you can tailor your engagement approach to address their specific concerns and needs."

III. Develop an engagement plan

Once you've identified your stakeholders, the next step is to develop a plan for engaging them at each stage of the implementation process. This could include a variety of communication channels and methods, such as town hall meetings, email updates, training sessions, and other communication methods.

For example, you might say something like, "Developing an engagement plan is critical to ensuring that stakeholders receive the information they need in a timely and effective manner. Consider using a variety of communication channels and methods to ensure that your message is heard. This could include town hall meetings, email updates, training sessions, and other communication methods."

IV. Communicate the vision

Communicating the vision of the change initiative is critical to engaging stakeholders effectively. Ensure that the communication is clear, concise, and tailored to each stakeholder group's needs and preferences. Explain how QBD will help the organization achieve its goals and objectives, and how stakeholder involvement is critical to realizing these benefits.

For example, you might say something like, "Communicating the vision of the change initiative is critical to ensuring that stakeholders understand the importance of the change and how it will benefit them. Use clear, concise language to explain how QBD will help the organization achieve its goals and objectives, and how stakeholder involvement is critical to realizing these benefits."

V. Create awareness

Creating awareness of the need for change and the benefits of QBD is essential to engaging stakeholders effectively. Use a variety of communication channels to ensure that stakeholders understand the importance of the change and how it will benefit them. This could include sharing success stories from other organizations that have implemented QBD, or demonstrating the potential benefits through case studies or simulations.

For example, you might say something like, "Creating awareness of the need for change and the benefits of QBD is essential to engaging stakeholders effectively. Use a variety of communication channels to ensure that stakeholders understand the importance of the change and how it will benefit them. This could include sharing success stories from other organizations that have implemented QBD, or demonstrating the potential benefits through case studies or simulations."

VI. Build desire

Building desire for the change initiative is critical to ensuring stakeholder buy-in and involvement. Use a variety of strategies to build desire, such as highlighting the potential benefits of QBD, providing incentives for involvement, and involving stakeholders in the planning and implementation process.

For example, you might say something like, "Building desire for the change initiative is critical to ensuring stakeholder buy-in and involvement. Consider highlighting the potential benefits of QBD, such as improved efficiency, enhanced productivity, and reduced costs. Providing incentives for involvement, such as recognition or rewards, can also be an effective way to build desire. Finally, involving stakeholders in the planning and implementation process can give them a sense of ownership and investment in the change initiative."

VII. Foster ownership

Fostering ownership of the change initiative is critical to ensuring stakeholder involvement and commitment. Involve stakeholders in the planning and implementation process, encourage feedback and input, and provide opportunities for stakeholders to take ownership of the change initiative.

For example, you might say something like, "Fostering ownership of the change initiative is critical to ensuring stakeholder involvement and commitment. Involve stakeholders in the planning and implementation process by forming working groups or committees, and encouraging feedback and input at every stage. Providing opportunities for stakeholders to take ownership of the change initiative, such as by leading training sessions or mentoring others, can also be an effective way to foster ownership."

VIII. Conclusion

Engaging stakeholders is critical to the success of any change initiative, and implementing QBD between groups within the same company is no exception. By identifying stakeholders, developing an engagement plan, communicating the vision, creating awareness, building desire, and fostering ownership, you can ensure that stakeholders are fully invested in the change initiative and that it has the best chance of success.

4. Engage stakeholders: Stakeholder engagement is critical for the success of the change initiative. Involve all relevant internal groups in the planning and implementation process.

This is when things really get exciting as we implement QBD between different groups in your company! Now, you may be thinking, "how can I get everyone on board with this change initiative?" The answer is simple: use the power of Appreciative Inquiry (AI) to engage stakeholders and build a sense of shared ownership and investment in the QBD implementation process.

Step 1: Identify the stakeholders

First things first, you need to identify the key stakeholders who will be impacted by the QBD implementation. You can't engage them if you don't know who they are! Use stakeholder mapping to identify the groups of people who will be impacted by the change initiative and their level of influence and interest in the initiative.

Step 2: Discover the strengths and successes

Now that you know who your stakeholders are, it's time to discover their strengths and past successes in QBD. This is where AI comes in! Conduct interviews or focus groups with employees or departments that have successfully implemented QBD in the past. Celebrate their successes and find out what worked well for them.

Step 3: Define the dream

Once you've identified the strengths and successes of your organization in QBD, it's time to define a clear and compelling vision for the QBD implementation. Involve all relevant internal groups in the process of defining the vision and make sure it's aligned with the overall goals and objectives of the organization. Remember, it's not just about the destination, it's about the journey too!

Step 4: Design the plan

With the vision in mind, it's time to design a plan for implementing the QBD methodology. Involve all relevant internal groups in the planning process and make sure the plan is actionable, realistic, and aligned with the overall goals and objectives of the organization. Don't forget to add some fun elements to the plan, like a QBD launch party or a QBD-themed escape room!

Step 5: Deploy the plan

Now it's time to deploy the QBD implementation plan and get everyone on board. Use a variety of engagement strategies such as town hall meetings, feedback sessions, and training sessions to make sure everyone is informed and involved in the implementation process. Remember, a spoonful of sugar helps the medicine go down! Find ways to make the implementation process fun and engaging, like a QBD-themed scavenger hunt.

Step 6: Discover and reinforce successes

Finally, continuously monitor and evaluate the QBD implementation using AI to discover and reinforce successes and build upon positive momentum. Provide recognition and rewards for successes and use them as a platform for building continued engagement and investment in the change initiative. Don't forget to celebrate your successes with a QBD-themed party or outing!

So, there you have it! With the power of AI and a little bit of creativity, you can engage stakeholders and make the QBD implementation process a fun and exciting journey for everyone involved. Remember, a happy team is a successful team!

5. Communicate effectively: Communicate the benefits of the new business methodology and how it will impact each internal group. Use multiple communication channels to ensure the message is received and understood.

I. Create a sense of urgency

To create a sense of urgency, communicate the need for change and the benefits of QBD. Use multiple communication channels, such as emails, town hall meetings, and one-on-one conversations to ensure that the message is received and understood. Use data to illustrate the urgency and emphasize the risks of not changing.

Example: "According to our analysis, we are losing $1 million a year due to inefficiencies in our production process. Implementing QBD can help us reduce these costs by 30%."

Quote: "Change almost never fails because it's too early. It almost always fails because it's too late." - Seth Godin

II. Form a powerful coalition

To form a powerful coalition, build a coalition of key stakeholders who support the implementation of QBD. Identify champions within each internal group and engage them in the planning and implementation process. Encourage open and honest communication to build trust and encourage collaboration.

Example: "We've identified a team of champions within each group who will help us plan and implement QBD. Their feedback and input will be critical to the success of this initiative."

Quote: "Alone we can do so little; together we can do so much." - Helen Keller

III. Create a vision for change

To create a vision for change, develop a clear vision of how the organization will look with QBD. Communicate the benefits of QBD and how it will impact each internal group. Use storytelling to help people visualize the future and build excitement.

Example: "With QBD, we will be able to reduce our production time by 50% and increase our profitability by 20%. We will be able to offer our customers better prices and faster delivery times."

Quote: "The visionary starts with a clean sheet of paper and reimagines the world." - Malcolm Gladwell

IV. Communicate the vision

To communicate the vision, use multiple communication channels. Use different methods to reach different internal groups, such as email updates, videos, and in-person meetings. Use storytelling and visual aids to make the message more memorable.

Example: "We will be launching a series of town hall meetings to communicate our vision for QBD. We will also be sending regular email updates and sharing videos to help people understand the benefits and impact of QBD."

Quote: "The single biggest problem with communication is the illusion that it has taken place." - George Bernard Shaw

V. Empower others to act on the vision

To empower others to act on the vision, provide resources and support. Provide training and coaching to help employees adjust to the new methodology. Encourage experimentation and provide a safe space for people to try new things.

Example: "We will be offering training and coaching to help people understand and implement QBD. We will also be setting up a 'sandbox' environment where people can experiment and test new ideas without fear of failure."

Quote: "The only way to do great work is to love what you do." - Steve Jobs

VI. Create short-term wins

To create short-term wins, celebrate milestones and progress made. Communicate the impact of QBD on each internal group and emphasize how it is contributing to the overall vision. Provide feedback and recognition to keep people motivated.

Example: "We have successfully reduced our production time by 25% in the first month of implementing QBD. This is a significant achievement and we want to thank everyone for their hard work and dedication."

Quote: "Success is not final, failure is not fatal: it is the courage to continue that counts." - Winston Churchill

VII. Consolidate gains and produce more change

After achieving short-term wins, it is important to consolidate gains and continue making progress. This involves integrating QBD into the organization's culture and providing ongoing support and training. Here are some suggestions on how to do this:

A. Make QBD a part of the organization's culture: To ensure that QBD becomes a part of the organization's culture, it is important to align it with the organization's values and goals. Develop communication strategies that highlight the positive impact of QBD and how it aligns with the organization's values.
B. Provide ongoing support and training: To ensure that QBD is adopted and integrated, provide ongoing support and training to employees at all levels of the organization. Develop a training program that includes refresher courses and hands-on training sessions to reinforce the benefits of QBD.
C. Develop metrics to track progress: Develop metrics to track progress towards achieving the goals and objectives identified in step II. Regularly provide updates to stakeholders, such as holding monthly meetings with managers to review progress and identify areas for improvement.
D. Celebrate successes: Celebrate successes and share them with the entire organization. This helps to build momentum and encourages employees to continue working towards achieving the goals and objectives.

For example, a manufacturing company implemented QBD to improve its production process. After achieving short-term wins, they developed a training program that included regular refresher courses and hands-on training sessions. They also established metrics to track progress and provided regular updates to stakeholders. As a result, they were able to integrate QBD into their culture and improve their production process.

VIII. Anchor new approaches in the organization's culture

Anchoring new approaches in the organization's culture involves making QBD a part of daily operations and ensuring that it is sustained. Here are some suggestions on how to do this:

A. Reinforce the benefits of QBD: Continually reinforce the benefits of QBD and how it aligns with the organization's values and goals. Use multiple communication channels to ensure that the message is received and understood.
B. Provide ongoing training and support: Provide ongoing training and support to employees to ensure that QBD is sustained. Develop a training program that includes hands-on training sessions and refresher courses.
C. Incorporate QBD into daily operations: Integrate QBD into daily operations by making it a part of the standard operating procedures. Regularly monitor and evaluate the implementation of QBD to ensure that it is sustained.
D. Provide opportunities for feedback: Provide opportunities for feedback from employees to identify areas for improvement and ensure that QBD is meeting the needs of the organization.

For example, a technology company implemented QBD to improve its software development process. They reinforced the benefits of QBD by regularly communicating how it aligns with the company's values and goals. They provided ongoing training and support to employees through a training program that included hands-on training sessions and refresher courses. They also incorporated QBD into daily operations by making it a part of their standard operating procedures. As a result, they were able to sustain the new approach and improve their software development process.

6. Train and support: Provide training and support to all internal groups to ensure they are equipped with the necessary knowledge and skills to effectively implement the new business methodology.

To achieve success, it's essential to provide training and support to all internal groups. By equipping employees with the necessary knowledge and skills, they can effectively implement the new business methodology. Let's dive into some practical steps for accomplishing this critical task.

I. Assess the Training Needs

The first step in providing training and support for QBD is to assess the training needs of each internal group. This involves identifying the specific skills and knowledge required to implement QBD effectively. For example, employees may need training in process mapping, value stream mapping, and statistical process control.

II. Develop a Training Plan

Once you've assessed the training needs of each internal group, the next step is to develop a training plan. This involves identifying the training resources required, such as trainers, training materials, and training facilities. It's also essential to determine the appropriate training methods and techniques. For example, hands-on training and coaching may be more effective than online training modules.

III. Deliver the Training

After developing a training plan, the next step is to deliver the training to all internal groups. It's essential to provide comprehensive training that covers all aspects of QBD, from basic concepts to advanced techniques. It's also important to ensure that the training is accessible to all employees, regardless of their location or job function. For example, online training modules may be more convenient for employees working remotely.

IV. Provide Ongoing Support

Training alone is not enough to ensure the successful implementation of QBD. It's also essential to provide ongoing support to employees to help them apply their new skills and knowledge effectively. This involves providing coaching, mentoring, and resources to help employees overcome any challenges they encounter during the implementation process.

V. Measure and Evaluate Training Effectiveness

Finally, it's crucial to measure and evaluate the effectiveness of the training provided. This involves identifying appropriate metrics to track training outcomes, such as employee satisfaction, knowledge retention, and application of new skills. It's also essential to evaluate the impact of training on the successful implementation of QBD.

Therefore, training and support are essential components of implementing QBD between groups within the same company. By assessing training needs, developing a training plan, delivering comprehensive training, providing ongoing support, and measuring training effectiveness, organizations can ensure that employees are equipped with the necessary knowledge and skills to implement QBD effectively. As Tom Peters once said, "Invest in training and development because it pays off in the long run."

7. Monitor and evaluate: Continuously monitor and evaluate the implementation of the new business methodology to identify any issues and make necessary adjustments.

Congratulations on successfully implementing QBD between teams in your company! But your work is not done yet. To ensure that the new business methodology is effective and sustainable, you must continuously monitor and evaluate its implementation to identify any issues and make necessary adjustments. Here's how you can accomplish this using the Lean methodology:

I. Establish Key Performance Indicators (KPIs)

To monitor the implementation of QBD, establish KPIs that measure the effectiveness of the new methodology. These KPIs should align with your goals and objectives and provide quantitative data on the performance of the new methodology. For example, a KPI for reducing process lead time could be the percentage of tasks completed on time.

II. Collect Data

Collect data regularly on the performance of the new methodology using the established KPIs. This data can be collected manually or automatically using software tools. Make sure that the data is accurate, complete, and reliable to make informed decisions.

III. Analyze Data

Analyze the collected data to identify areas where the implementation of QBD is not meeting the expected results. Use tools like Pareto charts and statistical process control to identify patterns and trends in the data. This will help you understand the root cause of the issues and make necessary adjustments.

IV. Take Action

Based on the data analysis, take action to make necessary adjustments to the implementation of QBD. Develop and implement solutions to address the root cause of the issues identified in the analysis. For example, if the KPI for reducing process lead time is not met, identify the bottlenecks in the process and find ways to remove them.

V. Monitor Progress

After taking action, monitor the progress of the implementation of QBD using the established KPIs. Collect data regularly to measure the impact of the adjustments made. Analyze the data and identify any additional issues that require further action.

VI. Standardize Best Practices

As you make adjustments to the implementation of QBD, identify the best practices that have been successful and standardize them. Ensure that everyone involved in the implementation of QBD is aware of these best practices and follows them.

By continuously monitoring and evaluating the implementation of QBD using the Lean methodology, you can identify any issues and make necessary adjustments to ensure the success and sustainability of the new business methodology. Remember, as W. Edwards Deming said, "In God we trust, all others must bring data."

Setting Out Targets to Attract Sources

In this chapter, we'll be diving into the important topic of "Setting out Targets to attract Sources." But before we get into the nitty-gritty details, let's take a moment to imagine a world where everything is haphazard and unorganized.

Picture this: You're standing in the middle of a crowded room, trying to reach a destination, but there are no signs or directions to guide you. You're surrounded by people, all rushing to get somewhere, but no one knows where they're going. Chaos reigns supreme.

Now imagine the opposite: You're standing in a spacious room with clear signs and directions pointing you towards your destination. You know exactly where you're going, and the journey is smooth and effortless. The same is true for businesses. Without clear targets and goals, a business can quickly become chaotic and directionless.

So, let's dive in and learn how to set targets that are specific, measurable, achievable, relevant, and time-bound (SMART), and align them with business goals and objectives. Get ready to take your business to the next level with the power of smart Target setting!

I. Identifying the right targets

A. Define your business goals and objectives

As a business owner or manager, one of the most important things you can do is to set goals and objectives for your company. These goals are critical to keeping your team focused and moving in the right direction. However, it's not just about setting any goals – it's about setting the right goals. That's where identifying the right targets comes in.

When it comes to identifying the right targets, there is a lot to consider. You need to think about your business's mission, values, and purpose. You need to think about the market you're in and the competition you face. You need to think about your customers and what they want. And you need to think about your resources and capabilities.

The best way to start identifying the right targets is by doing a SWOT analysis. This involves analyzing your company's strengths, weaknesses, opportunities, and threats. By understanding these factors, you can start to get a sense of what targets are realistic and achievable for your business.

Once you've done your SWOT analysis, it's time to start setting specific targets. This is where the SMART method comes in. SMART stands for specific, measurable, achievable, relevant, and time-bound. Let's take a closer look at each of these elements.

Specific: Your targets need to be specific. This means that they need to be clearly defined and focused. For example, instead of setting a target to "increase sales," you might set a target to "increase sales by 10% in the next quarter."

Measurable: Your targets need to be measurable. This means that you need to be able to track progress and determine whether you've achieved the target. For example, if your target is to increase sales by 10%, you need to be able to track sales and determine whether you've achieved that increase.

Achievable: Your targets need to be achievable. This means that they need to be realistic and within reach. For example, setting a target to increase sales by 1000% in the next quarter is not achievable for most businesses.

Relevant: Your targets need to be relevant to your business and aligned with your overall goals and objectives. For example, if your business is focused on sustainability, setting a target to reduce waste might be more relevant than setting a target to increase sales.

Time-bound: Your targets need to be time-bound. This means that you need to set a deadline for achieving the target. For example, setting a target to increase sales by 10% in the next quarter is time-bound because it has a deadline of the next quarter.

By setting SMART targets, you can ensure that you're focusing on the right things and that you have a clear plan for achieving your goals. In addition, by aligning your targets with your business goals and objectives, you can ensure that you're moving in the right direction and making progress towards your long-term vision.

That is why identifying the right targets is a critical part of setting goals and objectives for your business. By doing a SWOT analysis, setting specific, measurable, achievable, relevant, and time-bound targets, and aligning your targets with your overall business goals and objectives, you can ensure that you're setting the right goals and making progress towards your long-term vision.

B. Identify the areas of your business where improvement is needed

As a businessperson, it's important to identify the areas where your business needs improvement so you can focus your efforts on implementing solutions. One way to approach this is by utilizing Quantum Business Dynamics™ (QBD™) principles, which can help you identify areas that would benefit the most from process improvement.

Start by looking at your business as a whole and identifying any areas where there are bottlenecks, delays, or errors. These areas are likely to be good targets for process improvement since they are likely impacting your business efficiency and profitability. Use QBD principles to help you identify which areas are most critical, and prioritize accordingly.

To do this, start by defining the goals and objectives of your business, and identify the Key Performance Indicators (KPIs) that are most important to achieving those goals. This will help you identify the areas where improvement is needed, and prioritize which ones to tackle first.

Next, use data and metrics to identify where your business is falling short. This could include things like customer complaints, poor employee performance, or low sales figures. QBD principles can help you analyze this data and identify the root causes of these problems, so you can address them more effectively.

Another way to identify areas of improvement is by analyzing your business processes. Look for inefficiencies or redundancies in your workflow, and identify areas where you can streamline or automate tasks. QBD principles can help you map out your processes and identify areas where you can make improvements.

In addition to these methods, it can also be helpful to involve your employees in the process of identifying areas of improvement. They are often the ones who are most familiar with the day-to-day operations of your business and may have valuable insights to share. Use QBD principles to facilitate collaboration and feedback from your team members, and empower them to suggest solutions and improvements.

By using QBD principles to identify the areas of your business that need improvement, you can develop more effective solutions that are tailored to your unique needs and goals. With a clear understanding of where to focus your efforts, you can implement changes that will help your business become more efficient, profitable, and successful in the long run.

C. Analyze the performance data to determine where you need to set targets

Analyzing performance data is a critical aspect of determining where to set targets using QBD tools. Here are some proven business practices that can be used:

i. Review the data regularly: Set a regular schedule for reviewing your performance data, whether it be weekly, monthly, or quarterly. This will allow you to identify trends over time and pinpoint areas where improvements can be made.
ii. Identify the key performance indicators (KPIs): Determine the most important metrics that you should be tracking for your business. These may include things like revenue, customer satisfaction, and employee productivity. By focusing on these KPIs, you can gain a better understanding of where your business is excelling and where improvements are needed.
iii. Use data visualization tools: QBD tools often come with powerful data visualization capabilities that can help you quickly identify trends and patterns in your data. Use these tools to create charts, graphs, and other visualizations that will make it easier to understand your performance data at a glance.
iv. Compare your performance to industry benchmarks: Research industry benchmarks for your business and compare your performance against these benchmarks. This will give you a better understanding of how your business is performing relative to your peers and where you need to focus your efforts.
v. Analyze your data in context: Don't just look at your data in isolation. Instead, analyze it in the context of your business objectives and goals. This will help you identify areas where you need to improve in order to meet your objectives.

By applying these proven business practices, you can use QBD tools to analyze your performance data and determine where to set targets for your business. Remember to regularly review your data, identify KPIs, use data visualization tools, compare your performance to industry benchmarks, and analyze your data in context. With these techniques, you can improve your business performance and achieve your goals.

D. Identify the key performance indicators (KPIs) that will help you measure progress towards your targets

As a quality management methodology, QBD emphasizes the importance of identifying the right key performance indicators (KPIs) to measure progress towards your targets. KPIs are metrics that allow you to track how well your business is performing in specific areas, and they can be used to gauge whether you are meeting your targets.

To identify the right KPIs for your business, you need to start by looking at your goals and objectives. What are the specific outcomes you want to achieve, and what data points will give you insight into whether you are making progress towards those outcomes? It's important to make sure that the KPIs you choose are relevant to your business and aligned with your overall strategy.

Here are some general steps you can follow to identify the KPIs that will help you measure progress towards your targets using QBD:

i. Identify your business objectives and goals: Before you can identify KPIs, you need to have a clear understanding of what you are trying to achieve. This means setting specific, measurable, achievable, relevant, and time-bound (SMART) goals for your business.
ii. Determine the data you need to measure progress: Once you have identified your goals, you need to determine what data you need to measure progress towards those goals. This might include data from sales, marketing, customer service, finance, or other areas of your business.
iii. Choose the right KPIs: Once you have identified the data you need to measure progress, you can choose the KPIs that will help you track that data. Make sure that your KPIs are relevant to your goals and aligned with your overall strategy.
iv. Set benchmarks and targets: Once you have identified your KPIs, you need to set benchmarks and targets to help you measure progress. This means setting specific, measurable targets that you want to achieve over a given time period.
v. Track and analyze data: Once you have set your KPIs and targets, you need to track and analyze the data to see how well you are performing. This means using QBD tools to collect and analyze data, and using that data to make informed decisions about your business.

Overall, identifying the right KPIs is essential for measuring progress towards your targets using QBD. By following these steps, you can choose KPIs that are relevant to your business and aligned with your overall strategy, and use those KPIs to make informed decisions about your business.

II. Setting SMART targets

A. Be Clear

1. Define the specific outcome you want to achieve

Defining the specific outcome you want to achieve is a critical step in the QBD process. It is important to have a clear understanding of what you want to accomplish before you can set targets and begin measuring progress.

When defining your outcome, it is important to be specific and measurable. For example, if your goal is to increase sales, you should specify a target amount, such as a 10% increase in sales revenue over the next quarter. This target is specific and measurable, making it easier to track progress and determine whether you have achieved your goal.

Another important aspect of defining your outcome is to ensure that it aligns with your overall business goals and objectives. For example, if your business goal is to expand your market share, then your outcome should be related to increasing your customer base or increasing sales to existing customers. This will ensure that you are working towards your overall business goals.

In addition to being specific and aligned with your business goals, your outcome should also be achievable and relevant. It should be something that is realistic and within reach, while also being meaningful and relevant to your business.

Overall, defining the specific outcome you want to achieve is a crucial first step in the QBD process. By being specific, measurable, aligned with your business goals, achievable, and relevant, you can ensure that you are setting yourself up for success and making progress towards your overall objectives.

2. Identify the actions needed to achieve the outcome

To identify the actions needed to achieve the outcome, you can follow these steps:

a) Break down the outcome into smaller, measurable objectives that are specific and achievable.
b) Determine the tasks that need to be completed to achieve each objective.
c) Prioritize the tasks based on their importance and urgency.
d) Assign responsibilities for each task to the appropriate team member or department.
e) Set deadlines for each task and monitor progress regularly.
f) Identify any potential obstacles or challenges that may arise and

develop contingency plans to address them.

g) Ensure that resources, such as personnel, equipment, and budget, are allocated appropriately to support the actions needed to achieve the outcome.

By identifying the specific actions needed to achieve the outcome, you can create a roadmap that outlines the steps required to reach your goals. This will help you stay focused and organized, and ensure that your team is working towards the same objectives.

B. Measurability

1. Identify how you will measure progress towards the outcome

To measure progress towards the outcome, you will need to identify the specific key performance indicators (KPIs) that are relevant to your business and the outcome you want to achieve. These KPIs should be specific, measurable, and aligned with your business goals and objectives.

For example, if your outcome is to increase revenue, you may identify KPIs such as:

- Monthly sales revenue
- Customer acquisition rate
- Average order value
- Customer retention rate

Once you have identified your KPIs, you can set targets for each of them based on your desired outcome. You can then track your progress towards these targets using QBD tools such as dashboards and reports.

It is important to regularly review your KPIs and adjust your targets and actions as necessary. This will help ensure that you are making progress towards your desired outcome and that your business is on track to meet its goals.

2. Set quantitative or qualitative metrics that will help you measure success

When setting metrics to measure success towards an outcome, it is important to consider both quantitative and qualitative factors. Quantitative metrics involve measuring specific numerical values, such as revenue or profit margins, while qualitative metrics focus on non-numerical factors, such as customer satisfaction or employee morale.

To set effective metrics, it is important to ensure they are SMART: specific, measurable, achievable, relevant, and time-bound. Specific metrics should be clearly defined and align with the outcome you are trying to achieve. Measurable metrics should be easily quantifiable or trackable. Achievable metrics should be realistic and attainable within the desired timeframe. Relevant metrics should be directly related to the outcome and aligned with the overall business strategy. Time-bound metrics should have a specific timeframe for evaluation and adjustment.

Examples of quantitative metrics may include the number of units sold, website traffic, or profit margins. Examples of qualitative metrics may include customer satisfaction ratings, employee engagement scores, or social media sentiment analysis.

Ultimately, the metrics chosen will depend on the specific outcome and the nature of the business. It is important to regularly evaluate and adjust metrics to ensure they remain effective and aligned with business goals.

C. Achievability

1. Set targets that are challenging but realistic

Setting targets that are both challenging and realistic can be a delicate balance. On one hand, you want to push your team to achieve more than they may have thought possible. On the other hand, you don't want to set unrealistic goals that will only lead to frustration and burnout.

In implementing QBD, there are several strategies you can use to set targets that are both challenging and realistic:

a) Understand your team's capabilities: Before setting targets, it's important to have a clear understanding of your team's capabilities. Take the time to assess their skills, experience, and workload to determine what is realistically achievable. This will help you set targets that are challenging but still within reach.
b) Use data to inform your targets: QBD is all about using data to make informed decisions. When setting targets, use data to help you understand what is achievable and what is not. Look at past performance, industry benchmarks, and market trends to set targets that are both challenging and realistic.
c) Involve your team in the goal-setting process: When setting targets, it's important to involve your team in the process. This not only helps to ensure that the targets are realistic, but it also creates a sense of ownership and accountability among team members. Encourage your team to share their ideas and insights, and work together to set targets that everyone is invested in.
d) Break down larger goals into smaller, achievable targets: Sometimes, larger goals can seem overwhelming and unachievable. To make them more manageable, break them down into smaller, achievable targets.

This not only makes the goal more realistic, but it also provides a sense of progress and momentum as smaller targets are achieved along the way.

e) Celebrate successes along the way: When working towards challenging targets, it's important to celebrate successes along the way. This not only helps to keep morale high, but it also reinforces the idea that the targets are achievable. Celebrate milestones, recognize individual achievements, and take the time to acknowledge the hard work and effort that goes into achieving challenging targets.
f) Continuously evaluate and adjust targets as needed: QBD is an iterative process, and as such, it's important to continuously evaluate and adjust targets as needed. If targets are consistently missed, it may be time to re-evaluate their feasibility. Alternatively, if targets are consistently exceeded, it may be time to set more challenging targets to continue driving progress.

Setting targets that are both challenging and realistic is an important part of implementing QBD. By understanding your team's capabilities, using data to inform your targets, involving your team in the goal-setting process, breaking down larger goals, celebrating successes, and continuously evaluating and adjusting targets, you can set targets that push your team to achieve more while still being achievable.

2. Consider the resources needed to achieve the targets

When setting targets in QBD, it's essential to consider the resources needed to achieve them. Resources can include everything from financial resources to human resources, technology, and infrastructure. Without proper resources, achieving targets can be challenging or even impossible.

Here are some steps you can take to ensure you have the necessary resources to achieve your targets:

a) Assess your current resources: Before setting targets, it's essential to assess your current resources. This includes taking a close look at your budget, staff, and infrastructure. Determine if you have the resources needed to achieve the targets you're considering.
b) Identify any resource gaps: Once you've assessed your current resources, identify any resource gaps. These are the resources you don't have but need to achieve your targets. For example, you might need to hire additional staff or invest in new technology to achieve your targets.
c) Develop a plan to close resource gaps: Once you've identified resource gaps, develop a plan to close them. This might involve

reallocating existing resources, seeking funding or investments, or finding new partnerships. Whatever the plan, it's essential to ensure you have the necessary resources in place before setting targets.

d) Ensure resources are allocated appropriately: Once you have the necessary resources in place, ensure they are allocated appropriately. This means ensuring staff members are trained and equipped to achieve targets, infrastructure is updated and maintained, and financial resources are allocated appropriately.

e) Monitor resource allocation: Finally, monitor resource allocation throughout the target-setting process. Ensure resources are being used effectively and efficiently, and adjust your plan as needed to ensure targets are achieved.

By considering the resources needed to achieve targets, you can ensure that your QBD initiatives are successful. With the right resources in place, you can set challenging but realistic targets that drive growth and success for your organization.

D. Relevance

1. Ensure that the targets are aligned with your business goals and objectives

Aligning targets with business goals and objectives is a critical aspect of setting targets that are effective and relevant. This alignment ensures that the targets set are meaningful and contribute toward the overall success of the business.

To align targets with business goals and objectives, it is essential to identify the key priorities and strategic initiatives of the business. These priorities and initiatives should be based on the organization's mission, vision, and values.

Once these priorities and initiatives have been identified, the targets should be set in such a way that they contribute towards achieving these objectives. For instance, if the organization's goal is to increase revenue by a certain percentage, the targets set should be aligned with this goal, such as increasing sales or reducing costs.

In addition to aligning targets with business goals and objectives, it is essential to ensure that the targets set are specific, measurable, achievable, relevant, and time-bound (SMART). This framework ensures that the targets are realistic and attainable, making it easier to monitor progress and make adjustments where necessary.

It is also important to involve relevant stakeholders in the target-setting process. This involvement ensures that the targets set are relevant and meaningful to all stakeholders, increasing their commitment and motivation to achieve them.

Regularly monitoring progress towards the targets and making adjustments where necessary is also critical in ensuring that the targets set remain aligned with business goals and objectives. This monitoring helps identify any deviations and provides an opportunity to take corrective action to get back on track.

In summary, aligning targets with business goals and objectives is a crucial step in setting effective and relevant targets. It ensures that the targets set contribute towards achieving the organization's overall strategic objectives. To achieve this alignment, it is essential to involve relevant stakeholders, set SMART targets, and regularly monitor progress towards the targets.

2. Consider the impact of achieving the targets on your business

Considering the impact of achieving the targets on your business is an important step in setting realistic and effective targets using QBD. It is important to take a holistic approach and consider the potential impact on all aspects of the business, including financial, operational, and customer satisfaction.

Financial impact is one of the most important factors to consider when setting targets. This includes not only the potential revenue or cost savings that may be achieved through meeting the targets, but also the cost of implementing the strategies needed to reach those targets. It is important to determine if the resources needed to achieve the targets are financially feasible and if the potential benefits will outweigh the costs.

Operational impact is also important to consider. Meeting targets may require changes in processes or procedures, and it is important to determine if these changes are feasible and will not have a negative impact on other areas of the business. For example, setting a target for faster order fulfillment may require changes in production schedules or inventory management, and it is important to ensure that these changes do not negatively impact customer satisfaction or other operational areas.

Customer impact is another important factor to consider when setting targets. Targets that focus on improving customer satisfaction, such as reducing wait times or improving product quality, can have a significant impact on customer loyalty and retention. It is important to ensure that the targets are aligned with the needs and expectations of the target market and that achieving the targets will result in improved customer satisfaction.

In addition to these factors, it is also important to consider the impact of achieving the targets on the overall business strategy and objectives. The targets should be aligned with the overall mission and vision of the business, and achieving the targets should help move the business towards its long-term goals.

Overall, considering the impact of achieving the targets on the business is a critical step in setting realistic and effective targets using QBD. By taking a holistic approach and considering the potential financial, operational, and customer impact, as well as the alignment with business goals and objectives, businesses can set targets that are challenging but realistic and that will help drive growth and success.

E. Time-bound

1. Set a clear deadline for achieving the targets

Setting a clear deadline is an important step in ensuring that your targets are achieved in a timely manner. Without a deadline, there is no sense of urgency or accountability, which can result in targets being pushed back or not achieved at all.

When setting a deadline for achieving your targets, it is important to consider several factors:

1. Realistic time frame: The deadline should be realistic and achievable based on the available resources, capabilities, and scope of work. Setting an unrealistic deadline can lead to unnecessary stress and frustration, which can ultimately undermine your efforts.
2. Business needs: The deadline should align with your business needs and objectives. For example, if you are launching a new product, you may need to set a deadline that coincides with the launch date.
3. Flexibility: While it is important to have a clear deadline, it is also important to be flexible in the face of unforeseen circumstances. Build in some buffer time to account for unexpected delays or setbacks.
4. Tracking and monitoring: Once you have set a deadline, it is important to track and monitor progress regularly to ensure that you are on track to achieving your targets within the set time frame.
5. Communication: Communicate the deadline clearly with all stakeholders and team members involved in achieving the targets. This will help ensure that everyone is aware of the timeline and can work towards the same goal.

Setting a clear deadline can help create a sense of urgency and focus, which can ultimately lead to greater productivity and achievement of targets. By taking the time to consider the factors above and communicating clearly with stakeholders, you can set realistic and achievable targets that are aligned with your business goals and objectives.

2. Break down the targets into smaller, more manageable milestones

In QBD, it is essential to break down targets into smaller, more manageable milestones. By doing so, you can measure progress towards your goals and make any necessary adjustments along the way. Breaking down targets into smaller milestones also helps to ensure that the project stays on track and does not become overwhelming.

To implement QBD effectively, you should identify key performance indicators (KPIs) that will help you measure progress towards your targets. From there, you can set quantitative or qualitative metrics that will help you measure success. It's also crucial to set targets that are challenging but realistic and consider the resources needed to achieve the targets.

To align your targets with your business goals and objectives, you need to define the specific outcome you want to achieve and identify the actions needed to achieve that outcome. You must consider the impact of achieving the targets on your business and set a clear deadline for achieving the targets.

Breaking down targets into smaller milestones also ensures that the targets are achievable, and it helps to keep the team motivated. It's easier to measure progress towards a small goal, which provides a sense of accomplishment and progress.

With QBD, you can break down targets into manageable milestones, set achievable targets, measure progress towards those targets, and align your targets with your business goals and objectives. By doing so, you can stay on track, adjust your approach when necessary, and ultimately achieve success in your business.

III. Aligning targets with business goals and objectives

A. Ensure that targets are aligned with the company's vision and mission

When setting targets for your company using QBD, it is important to ensure that they are aligned with your company's vision and mission. This means that your targets should be consistent with your overall business strategy and should be designed to support the long-term success of your company.

To ensure that your targets are aligned with your company's vision and mission, start by reviewing your company's strategic plan. This plan should outline your company's goals and objectives, as well as the strategies you will use to achieve them. Use this plan as a guide when setting your targets to ensure that they support your overall business strategy.

It is also important to consider how each target fits into the bigger picture. Make sure that your targets are not isolated goals, but rather are part of a larger plan that supports your overall business objectives. This will help ensure that your efforts are focused on achieving your long-term goals rather than just short-term gains.

Another key consideration when setting targets is to ensure that they are realistic and achievable. While it is important to set challenging targets that push you and your team to excel, it is equally important to ensure that they are achievable given the resources and time available. Setting unrealistic targets can lead to frustration and demotivation, which can ultimately hurt your chances of achieving your goals.

One way to ensure that your targets are both challenging and achievable is to break them down into smaller, more manageable milestones. This will help you track your progress and adjust as needed to keep you on track to achieve your targets.

In addition to ensuring that your targets are aligned with your company's vision and mission, it is also important to consider the impact of achieving these targets on your business. Will achieving these targets help to increase revenue or reduce costs? Will it improve customer satisfaction or employee engagement? Understanding the potential impact of achieving your targets can help you prioritize them and allocate resources accordingly.

Finally, set clear deadlines for achieving your targets. This will help ensure that your team is focused and motivated to achieve their goals, and will help you track progress and make adjustments as needed to stay on track. By setting targets that are aligned with your company's vision and mission, realistic and achievable, and broken down into smaller milestones, you can help ensure that you are on the path to long-term success.

B. Consider how the targets will impact other areas of the business

When setting targets for a specific area of the business, it is important to consider how those targets may impact other areas of the business. Targets that may seem beneficial for one area of the business could potentially have negative effects on other areas. For example, a target to increase sales on a particular product line may put a strain on the company's production resources, which could affect the company's ability to meet demand for other products.

This is where QBD can be useful in ensuring that targets are aligned with the broader goals and objectives of the company. By using QBD to assess the impact of the targets on other areas of the business, companies can make informed decisions and prioritize targets that will have the most positive impact overall.

One approach to considering the impact of targets on other areas of the business is to use a SWOT analysis. This involves analyzing the company's strengths, weaknesses, opportunities, and threats in relation to the targets being considered. By conducting a SWOT analysis, companies can identify potential conflicts or unintended consequences of the targets and adjust them accordingly.

Another approach is to involve stakeholders from across the business in the target-setting process. By soliciting input from departments and teams that may be affected by the targets, companies can ensure that all perspectives are taken into account and potential conflicts are identified early on. This can also help to build buy-in and support for the targets across the business.

In summary, when setting targets for a specific area of the business, it is important to consider the potential impact on other areas of the business. Using QBD to assess the impact of targets and involving stakeholders from across the business can help to ensure that targets are aligned with the company's broader goals and objectives and avoid unintended consequences.

C. Involve stakeholders in the process of setting targets to ensure buy-in and support

Involving stakeholders in the process of setting targets can be a critical step towards achieving them. By involving key stakeholders, you can gain support for your targets and ensure that they are aligned with the needs and priorities of your organization. Here are some tips for involving stakeholders in the target setting process:

1. Identify your stakeholders: Identify the individuals or groups that will be impacted by the targets you are setting. This may include employees, customers, investors, and other partners.
2. Communicate clearly: Once you have identified your stakeholders, it is important to communicate clearly with them about the targets you are setting. Explain the rationale behind the targets and how they will benefit the organization as a whole.
3. Solicit feedback: Encourage your stakeholders to provide feedback on the targets you are setting. This can help you identify potential issues or challenges that may arise and address them before they become a problem.
4. Get buy-in: Once you have set your targets, it is important to get buy-in from your stakeholders. This can be done by communicating the targets clearly and explaining how they align with the needs and priorities of the organization.
5. Hold stakeholders accountable: Finally, it is important to hold your stakeholders accountable for supporting the targets you have set. This can be done by providing regular updates on progress towards the targets and holding individuals and teams accountable for achieving their respective targets.

By involving stakeholders in the target setting process, you can ensure that your targets are aligned with the needs and priorities of your organization, gain support for your targets, and hold individuals and teams accountable for achieving their respective targets.

D. Communicate the targets and their importance to all relevant stakeholders

Communicating targets and their importance to relevant stakeholders is a critical step in ensuring successful implementation of a quality-by-design (QBD™) approach. It is essential to involve stakeholders in the process of setting targets to ensure buy-in and support. Once the targets have been set, communicating them effectively is crucial to ensuring that everyone understands the expectations and their role in achieving them.

To communicate targets effectively, it is important to consider who the stakeholders are and how best to reach them. This may involve using different methods for different stakeholders, such as in-person meetings, emails, newsletters, or social media. It is also important to use clear and concise language and to provide examples or visual aids to help stakeholders understand the targets and their importance.

When communicating the targets, it is important to emphasize how they align with the company's vision and mission. This helps stakeholders understand the bigger picture and how their efforts are contributing to the company's overall success. Additionally, it is important to explain how achieving the targets will benefit stakeholders individually, as well as the company as a whole. This helps to create a sense of ownership and motivation among stakeholders to work towards the targets.

Another important aspect of communicating targets is to provide regular updates on progress towards achieving them. This helps stakeholders understand the progress being made, areas where improvements are needed, and how their efforts are contributing to achieving the targets. It also helps to keep stakeholders motivated and engaged in the process.

Thus, communication is a critical factor in the successful implementation of a QBD approach. By involving stakeholders in the process of setting targets and communicating them effectively, companies can create a shared vision for success and ensure that everyone is working towards the same goals.

E. Monitor progress towards the targets and adjust as needed to stay aligned with business goals and objectives.

Monitoring progress towards targets is crucial to ensure that the business stays on track and achieves its objectives. Here are some steps to effectively monitor progress:

1. Establish a reporting system: Set up a system that enables you to track and report progress towards your targets. This could be a dashboard, spreadsheet, or other tool that allows you to monitor and track key metrics.
2. Assign responsibility: Identify who will be responsible for monitoring progress towards the targets. This person should have the authority and resources necessary to adjust as needed.
3. Set regular checkpoints: Schedule regular checkpoints to review progress towards the targets. This could be a weekly, monthly, or quarterly meeting to discuss progress and make any necessary adjustments.
4. Analyze data: Use the data you have collected to analyze progress towards your targets. Look for trends and patterns that can help you identify areas where you are making progress and areas where you need to make adjustments.
5. Adjust: Based on your analysis, make any necessary adjustments to your targets, processes, or resources. This could involve increasing resources in areas where progress is lagging or adjusting targets that are unrealistic or not aligned with business goals.
6. Communicate progress: Keep stakeholders informed of progress towards the targets. This could involve regular updates via email, a dashboard, or other means of communication.
7. Celebrate success: When you reach your targets, take time to celebrate your success. Recognize the hard work that went into achieving the targets and reward those who contributed to the success.

By monitoring progress towards your targets and making adjustments as needed, you can ensure that your business stays on track and achieves its goals and objectives.

Setting out targets that are specific, measurable, achievable, relevant, and time-bound is a crucial step towards achieving success in any business venture. By using the tools provided by QBD, businesses can ensure that their targets are aligned with their vision and mission, and that they are challenging but achievable. It is important to involve stakeholders in the target-setting process to ensure buy-in and support, and to communicate the targets and their importance to all relevant parties. By monitoring progress towards the targets and adjusting as needed, businesses can stay aligned with their goals and objectives.

However, setting targets is just one part of the equation. In the next section, we will explore how businesses can bridge themselves to other businesses to achieve even greater success. By forming strategic partnerships and collaborations, businesses can tap into new markets, gain access to valuable resources and expertise, and build mutually beneficial relationships with other organizations. So, let's move on to the next section and discover how to bridge your business to other businesses for even greater success.

Your Business Bridged to Other Businesses

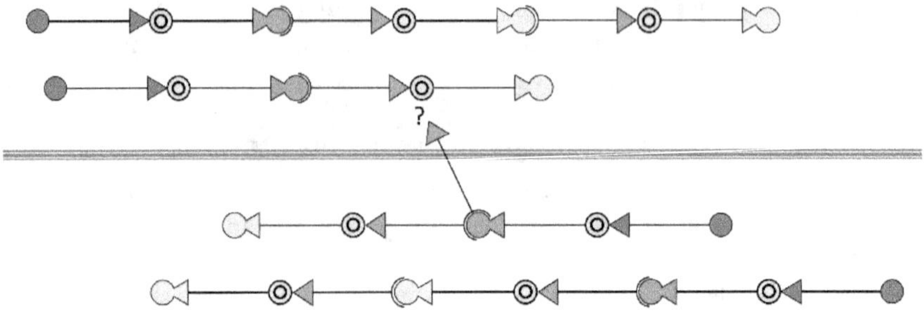

I. Introduction

A. Importance of partnerships in business

Partnerships are a key component of success in modern business. As the business landscape becomes more complex and competitive, companies that forge partnerships with other organizations can leverage their strengths to gain a competitive advantage. This advantage can take many forms, from sharing resources and expertise to accessing new markets and customers.

One of the most significant benefits of partnerships is the ability to pool resources. By working with another organization, businesses can share the costs of research and development, marketing, and other expenses, allowing them to invest more in their core competencies. This sharing of resources can also lead to greater innovation, as partners bring together different perspectives and approaches to problem-solving.

Partnerships can also provide access to new markets and customers. For example, a company that partners with a larger, more established organization can tap into their customer base and distribution network, helping them to reach new audiences and increase their market share. Similarly, a company that partners with a smaller, more nimble organization can gain access to emerging markets and new technologies.

Another benefit of partnerships is the ability to share expertise. By working with other organizations, businesses can tap into their partners' knowledge and experience, gaining insights and expertise that they may not have in-house. This can lead to greater efficiency, better decision-making, and improved outcomes.

However, partnerships are not without their challenges. For example, businesses must carefully consider the potential risks and rewards of partnering with another organization. They must also ensure that their partners share their values and goals, and that they have compatible cultures and working styles.

Another challenge is the need for effective communication and collaboration. Partnerships require a high degree of trust and openness, and businesses must be willing to share information and work together closely to achieve their goals. They must also be willing to compromise and adapt their strategies to fit their partners' needs and objectives.

Despite these challenges, the benefits of partnerships far outweigh the costs. By working together, businesses can achieve more than they could on their own, and they can do so in a way that is mutually beneficial and sustainable over the long term.

In the context of QBD, partnerships are critical to achieving success in a highly interconnected and complex business environment. QBD recognizes that businesses are not isolated entities, but rather interconnected nodes in a vast network of relationships and transactions. By leveraging the power of partnerships, businesses can create more robust and resilient networks that are better equipped to adapt to changing market conditions and emerging threats.

QBD also recognizes that partnerships are not one-size-fits-all. Each partnership must be carefully crafted to fit the specific needs and objectives of the partners involved. QBD provides a framework for creating these partnerships, identifying the key players and their roles, and establishing the rules and incentives that will govern their interactions.

Ultimately, QBD's focus on partnerships reflects a broader shift in the business world towards collaboration and co-creation. As companies recognize the benefits of working together, they are increasingly looking for ways to build partnerships that will help them achieve their goals and stay ahead of the competition. By embracing this new paradigm, businesses can unlock new opportunities and create a more sustainable and prosperous future for themselves and their partners.

B. Overview of bridging methodology

Overview of Bridging Methodology

Bridging methodology is a process of identifying, designing, and implementing connections between different business entities, creating opportunities for new revenue streams, enhanced customer experiences, and optimized business processes. This approach is driven by the idea that businesses can achieve greater success and growth by collaborating with others instead of working in isolation.

At the core of bridging methodology is the concept of identifying and leveraging the unique strengths, capabilities, and assets of each business to create new value for all parties involved. By understanding the capabilities and value propositions of other businesses, companies can identify opportunities to collaborate, share resources, and co-create new products or services.

One classic example of how bridging methodology can be applied is in the hospitality industry. In this industry, businesses such as hotels, resorts, and travel agencies can benefit from forming partnerships with other businesses such as airlines, transportation companies, and local tourist attractions. By doing so, they can create a seamless and integrated travel experience for their customers, improving customer satisfaction and increasing repeat business.

For instance, a hotel could partner with a local transportation company to offer shuttle services to and from the airport or train station, making it easier for guests to get to their accommodations. The hotel could also partner with a nearby tourist attraction to offer discounted tickets to guests, providing an added incentive for customers to choose that particular hotel over others.

Furthermore, the hotel could partner with a travel agency to create a customized travel package that includes airfare, accommodations, transportation, and tickets to local attractions, providing customers with an all-inclusive travel experience. This partnership could be mutually beneficial, as the travel agency could leverage the hotel's customer base to promote their services, while the hotel could benefit from increased bookings and revenue.

By adopting a bridging methodology, businesses can create a network of partners that help them enhance their offerings, create new value propositions, and optimize their operations. Moreover, this approach allows businesses to tap into new markets, reach new customer segments, and build long-lasting relationships that drive growth and innovation.

The bridging methodology also provides a framework for businesses to identify and prioritize their partnerships based on their strategic objectives and core competencies. By understanding their own strengths and capabilities, businesses can identify areas where they can add the most value to potential partners and target businesses that complement their offerings.

For instance, a hotel that specializes in luxury accommodations could target high-end travel agencies or local tourist attractions that cater to affluent customers. By doing so, they could create a premium travel experience that aligns with their brand image and attracts customers who value luxury and exclusivity.

In summary, the bridging methodology is a powerful approach that enables businesses to create value by collaborating with others. By identifying and leveraging the unique strengths and capabilities of each partner, businesses can create new revenue streams, improve customer experiences, and optimize their operations. The hospitality industry is just one example of how this methodology can be applied, but the possibilities for bridging are endless across all industries. As businesses continue to adopt this approach, they will be able to unlock new opportunities, achieve greater success, and drive innovation in their respective markets.

C. Benefits of using QBD for business partnerships

Business partnerships have always been important for businesses seeking to expand their reach, increase their profitability, and stay competitive in a crowded market. However, traditional partnership models often rely on cumbersome and manual processes, leading to delays and errors that can be costly and time-consuming. This is where QBD comes in, offering a seamless and efficient way to bridge businesses and reap the benefits of partnership.

Let's take the example of a hospitality business seeking to partner with local tour operators to provide unique experiences to its guests. Using QBD, the hospitality business can set up a bridge between their booking system and the tour operators' reservation system. This allows for real-time availability and pricing information to be shared between the two systems, streamlining the booking process for guests and reducing the risk of double bookings or errors.

But QBD goes beyond just connecting two systems - it also allows for the creation of Bi-link tokens that represent the value exchanged between the two businesses. In this example, each completed booking results in a Bi-link token that represents the value of the guest's experience, with one token going to the hospitality business and one going to the tour operator. These tokens can be used to track and incentivize partnership success, with bonuses or rewards offered for reaching certain milestones or achieving specific targets.

Another benefit of using QBD for business partnerships is the ability to easily integrate and collaborate with other partners. In the hospitality example, this could mean expanding the partnership network to include restaurants, transportation providers, and other local businesses. QBD allows for the creation of a flexible and dynamic ecosystem of partners, each contributing unique value and benefiting from the value exchanged.

Furthermore, QBD provides an added layer of security and transparency to business partnerships. The use of Bi-link tokens ensures that all transactions are recorded and validated, reducing the risk of fraud or disputes. Additionally, the ability to view and track the flow of value exchanged between partners allows for a greater understanding of the partnership's overall success and areas for improvement.

Overall, the benefits of using QBD for business partnerships are numerous and far-reaching. By streamlining processes, creating a dynamic ecosystem of partners, and providing added security and transparency, QBD can help businesses achieve their partnership goals and stay competitive in a constantly evolving market.

II. Identifying Potential Targets

A. Market research and analysis

Once a business has implemented a Bi-link chain solution for their internal processes, the next step is to start expanding that chain by bridging to other businesses. One way to identify potential Targets for bridging is through market research and analysis.

Market research is a critical tool for any business looking to expand and identify potential Targets for partnership. Traditional market research techniques include surveys, focus groups, and analyzing customer data. However, with the implementation of QBD, businesses can take their market research to the next level.

By bridging with other businesses in the same industry, a company can gain valuable insights into the market trends, customer behavior, and other key performance indicators. This not only helps in identifying potential Targets for bridging, but it can also help in fine-tuning their own internal processes.

For example, let's revisit the hospitality use case we discussed earlier. A hotel chain that has implemented QBD for their internal Bi-link chain solution can bridge to other businesses in the hospitality industry to gather data on market trends and customer behavior. By partnering with online travel agencies (OTAs), for instance, the hotel chain can gain access to data on customer preferences, booking patterns, and other valuable information that can be used to optimize their own booking process and tailor their offerings to the needs of their customers.

Furthermore, bridging with other businesses in the same industry can also lead to valuable collaborations and partnerships. By sharing data and insights, businesses can work together to develop new products and services that better serve the needs of their customers. For example, a hotel chain can partner with a car rental company to offer customers bundled packages that include both hotel accommodations and rental cars, providing customers with a one-stop-shop for their travel needs.

QBD can also be used to bridge with businesses outside of the hospitality industry for market research and analysis. For instance, a hotel chain can bridge with a ride-sharing company to gather data on the travel patterns of their customers. This information can then be used to optimize the hotel's location and transportation offerings, and even develop new services that cater to their customers' needs.

QBD can greatly benefit businesses in identifying potential Targets for bridging and expanding their Bi-link chain solutions. By using QBD for market research and analysis, businesses can gain valuable insights into market trends, customer behavior, and other key performance indicators. This not only helps in identifying potential Targets for partnership, but it can also help in fine-tuning their own internal processes and developing new products and services that better serve the needs of their customers.

B. Identifying complementary businesses

An executive using QBD to identify complementary businesses should start by examining their own business and identifying the products, services, and processes that are critical to their success. They should then consider the ecosystem of businesses that support and rely on those critical elements. For example, a hotel might identify its key success factors as high occupancy rates, exceptional customer service, and efficient housekeeping operations. Complementary businesses might include travel agencies, transportation services, and housekeeping supply companies.

Once complementary businesses have been identified, the executive can use QBD to identify potential Targets for bridging. They should consider factors such as the target's size, reputation, and expertise, as well as the potential for a mutually beneficial relationship. The executive can use QBD to analyze the target's business processes, identify areas where bridging could improve efficiency or enhance the customer experience, and propose a Bi-link chain solution that benefits both businesses.

Market research and analysis can also play a crucial role in identifying complementary businesses. The executive can use QBD to gather data on industry trends, customer preferences, and the competitive landscape, and use that information to identify potential Targets for bridging. For example, a hotel might analyze data on customer travel patterns to identify popular destinations and transportation services, and use that information to identify complementary businesses to bridge with.

Another important tool for identifying complementary businesses is networking. The executive can use QBD to map out their existing network of business contacts, and use those connections to identify potential Targets for bridging. They can also attend industry events, join professional organizations, and participate in online forums to expand their network and identify new opportunities for bridging.

Ultimately, the key to identifying complementary businesses is to approach the process with an open mind and a willingness to think creatively. QBD provides a powerful framework for analyzing business processes, identifying potential Targets for bridging, and proposing innovative Bi-link chain solutions that benefit both businesses. By leveraging the power of QBD, executives can identify complementary businesses, create new revenue streams, and drive business growth in a rapidly changing marketplace.

C. Prioritizing potential Targets

There are several ways in which business leaders can prioritize potential Targets when implementing QBD solutions with the help of Data Stewardship Solutions consulting and QBD tools and methodology. Here are some approaches that can be used:

1. Analyze business goals and objectives: The first step is to identify the business goals and objectives that need to be achieved through partnerships. This can be done by looking at the company's overall strategy, growth plans, and revenue targets. Once the goals and objectives are identified, the next step is to prioritize potential Targets based on their ability to help achieve those goals.
2. Evaluate the potential impact of the partnership: The next step is to evaluate the potential impact of the partnership on the company's business. This can be done by analyzing factors such as the size of the market, the potential revenue generated, and the level of competition. By evaluating the potential impact of the partnership, business leaders can prioritize potential Targets that are most likely to have a significant impact on their business.
3. Assess the level of compatibility: It's important to assess the level of compatibility between the two companies before entering into a partnership. This can be done by analyzing factors such as the company's culture, values, and business model. By prioritizing potential Targets that are most compatible with the company's values and business model, business leaders can increase the chances of a successful partnership.
4. Use QBD tools and methodology: QBD tools and methodology can be used to identify potential Targets based on various criteria such as market size, revenue potential, and level of competition. By using these tools, business leaders can prioritize potential Targets that are most likely to result in a successful partnership.
5. Leverage AI as a service: AI can be used to analyze large amounts of data to identify potential Targets that are most likely to result in a successful partnership. By leveraging AI as a service, business leaders can prioritize potential Targets based on various criteria such as market trends, customer behavior, and competitive landscape.

By using a combination of these approaches, business leaders can effectively prioritize potential Targets and increase the chances of a successful partnership.

III. Creating Bridges

A. Determining value exchanges

Determining value exchanges is an essential step in creating successful business partnerships. It involves identifying what each business brings to the table and what they stand to gain from the partnership. Value exchange is the cornerstone of a QBD approach, and it is essential to create a mutual win-win situation for all parties involved.

One approach to determining value exchanges is to conduct a thorough analysis of both businesses' strengths, weaknesses, opportunities, and threats (SWOT analysis). This analysis helps identify areas of complementarity between the businesses and highlights areas where they can work together to achieve common goals. SWOT analysis is an excellent tool for determining the value that each business can bring to the partnership.

Another approach is to look at the different stages of the value chain and identify areas where the two businesses can collaborate. For instance, in the hospitality industry, the hotel may be focused on providing comfortable accommodation to guests, while the restaurant may be focused on providing excellent dining experiences. By collaborating, the hotel and the restaurant can create an integrated offering that provides guests with a complete experience. The hotel can refer guests to the restaurant, and the restaurant can offer discounts to hotel guests, resulting in a mutually beneficial partnership.

Another technique for determining value exchanges is to look at the different capabilities that each business brings to the partnership. This approach requires identifying the unique strengths of each business and determining how those strengths can be leveraged to create value. For example, in the hospitality industry, a hotel may have an extensive customer database, while a restaurant may have a unique cuisine that attracts a particular type of guest. By collaborating, the hotel and the restaurant can share their capabilities to create targeted marketing campaigns that appeal to specific customer segments.

A further approach is to consider the potential risks involved in the partnership and determine how each business can mitigate those risks. This approach requires businesses to think critically about potential challenges that could arise and develop contingency plans to overcome them. This approach can help businesses create value by ensuring that the partnership is resilient and can withstand any challenges that may arise.

Lastly, it's crucial to consider the cultural fit between the businesses.

Businesses with similar values and cultures are more likely to create successful partnerships than those that are fundamentally different. By considering cultural fit, businesses can create partnerships that are built on trust, respect, and a shared vision for success.

To sum up, determining value exchanges is essential to creating successful business partnerships. Businesses should consider their strengths, weaknesses, opportunities, and threats, look at the different stages of the value chain, identify unique capabilities, consider potential risks, and ensure cultural fit. With these techniques, businesses can create partnerships that are built on mutual benefits, trust, and respect.

B. Negotiating terms and agreements

After identifying the potential Targets and determining the value exchanges that can be made, the next step in bridging your business to other businesses is to negotiate the terms and agreements of the partnership. This can be a delicate process, as both parties will likely have their own goals, priorities, and concerns to consider.

The first step in negotiating terms and agreements is to clearly communicate the value proposition that each party brings to the partnership. This should include a detailed explanation of the benefits that each party will receive, such as increased revenue, access to new markets, or enhanced capabilities. It is also important to discuss any risks or potential challenges associated with the partnership, as well as any contingencies that may need to be put in place to mitigate those risks.

Once both parties have a clear understanding of the value that can be created through the partnership, the next step is to negotiate the specific terms and conditions of the agreement. This may include discussions around pricing, revenue sharing, intellectual property rights, data ownership, and other legal and financial considerations.

One approach to negotiating terms and agreements is to focus on creating a win-win scenario for both parties. This means looking for areas where both parties can benefit, and finding ways to structure the agreement in a way that is mutually beneficial. This may involve making compromises or concessions in certain areas, while still ensuring that both parties are receiving the value they need to justify the partnership.

Another important consideration when negotiating terms and agreements is to ensure that the agreement is scalable and flexible enough to accommodate future changes and growth. This may involve building in provisions for expansion, termination, or modification of the agreement, as well as establishing clear communication channels and reporting mechanisms to ensure that both parties are staying informed and up to date on the status of the partnership.

Finally, it is important to ensure that the terms and agreements of the partnership are well-documented and legally binding. This may involve working with legal experts and other advisors to ensure that all aspects of the agreement are in compliance with relevant laws and regulations, as well as ensuring that both parties have a clear understanding of their rights and obligations under the agreement.

Negotiating terms and agreements is a critical step in bridging your business to other businesses. By clearly communicating the value proposition, focusing on creating win-win scenarios, building in scalability and flexibility, and ensuring that the agreement is well-documented and legally binding, businesses can establish successful partnerships that can drive growth, innovation, and value creation for all parties involved.

C. Establishing bridging protocols

Establishing bridging protocols is a critical aspect of successful business partnerships. It provides clear guidelines and expectations for both parties, reducing the risk of misunderstandings and disputes. Here are some solid and helpful suggestions for establishing bridging protocols:

1. Start with clear communication: Before any partnership begins, both parties must be open and transparent about their goals, expectations, and limitations. This communication should continue throughout the entire partnership to ensure that both parties remain on the same page.
2. Define roles and responsibilities: Clearly defining the roles and responsibilities of each party is critical to the success of any partnership. This will help ensure that everyone knows what is expected of them, and no one is left guessing.
3. Establish protocols for data sharing: With QBD, data sharing is essential for creating bridges between businesses. However, this can also pose risks to both parties' data security and privacy. Therefore, establishing protocols for data sharing is crucial to ensure that both parties' data is protected.
4. Establish protocols for conflict resolution: Even the best partnerships may encounter conflicts or misunderstandings along the way. Establishing protocols for conflict resolution can help prevent these conflicts from escalating and ensure that they are resolved in a timely and fair manner.
5. Establish protocols for monitoring and reporting: QBD relies on real-time monitoring and analysis to ensure that the Bridges are functioning as intended. Establishing protocols for monitoring and reporting can help identify and address issues quickly, preventing any significant disruptions to the partnership.
6. Include exit strategies: While no one wants to consider the possibility of

a partnership ending, it is essential to include exit strategies in the bridging protocols. This ensures that both parties have a clear understanding of how the partnership can be dissolved if necessary and what the consequences of such dissolution may be.

7. Establish protocols for ongoing evaluation: Partnerships evolve over time, and it is essential to regularly evaluate their effectiveness. Establishing protocols for ongoing evaluation can help identify areas for improvement and ensure that the partnership continues to provide value for both parties.

In summary, establishing bridging protocols requires clear communication, defining roles and responsibilities, protocols for data sharing, conflict resolution, monitoring and reporting, exit strategies, and ongoing evaluation. By following these suggestions, businesses can establish strong and successful partnerships that provide mutual benefits for all parties involved.

IV. Implementing Bridges

A. Integrating QBD systems

Integrating QBD systems can seem like a daunting task, but with the right approach, it can be a seamless and efficient process. Implementing Bridges in QBD requires a thoughtful and strategic approach to ensure that both businesses are working towards the same goal and that the integration is successful.

The first step in integrating QBD systems is to identify the specific business process that needs to be integrated. This process could be anything from order management to supply chain management or data sharing. Once the process is identified, the next step is to determine which data is required for the process and what data will be shared between the two businesses.

QBD's API makes integration with existing platforms and processes a breeze. Our human control panels further simplify the process, making it easy to maximize existing capital outlays. Our AI seed data helps clients easily create stubs of code that can integrate with existing systems without the need for exposing their own IP to outside consultants. Using OpenAPI, our APIs provides easy documentation and integration with most languages, making it easy for businesses to integrate with QBD.

Once the data is identified, the next step is to determine the protocols and security measures that will be used for data exchange. QBD provides high-level security protocols such as Sig 4 and libsodium to promote secure data exchanges. Our trustless implementation of architecture ensures that data is secure and that it is impossible to decrypt. One-way hashing further ensures data privacy.

Conslice is another service provided by QBD that provides a local slice of the consent mechanism for QBD. This service allows data to remain within a business facility while providing consent for high-end computing such as Big Data map reductions as a mathematical component of map reduce. This saves businesses costly data conditioning such as "scrubbing data" while keeping them compliant with all 238 privacy regulations.

The final step in integrating QBD systems is to create a plan for testing and implementation. This plan should outline the testing process, including what data will be used, who will perform the tests, and what metrics will be used to determine success. Once testing is complete, the implementation process can begin, and data can be shared between the two businesses.

Overall, integrating QBD systems is a straightforward process that requires careful planning and execution. With QBD's API and security protocols, businesses can integrate their systems and share data with ease. By following the steps outlined above, businesses can successfully implement Bridges in QBD, which can lead to increased revenue, improved efficiency, and better customer experiences.

B. Ensuring data security and privacy

As businesses continue to implement QBD systems and create bridges with other businesses, it is essential to ensure the security and privacy of the data being exchanged. Data breaches can cause significant harm to businesses, not only in terms of financial loss but also in terms of damage to reputation and loss of customer trust. In this section, we will discuss the importance of data security and privacy in QBD, and provide some best practices for ensuring the safety of your data.

One of the main benefits of QBD is the ability to securely exchange data between businesses, even if they have different security protocols and systems in place. However, this does not mean that businesses should neglect their own security measures. Each business must ensure that their own data is secure before engaging in QBD.

Businesses can start by implementing strong access controls and permissions for their data. This means that only authorized individuals have access to sensitive data, and that access is granted based on a need-to-know basis. Additionally, businesses should implement data encryption to ensure that even if data is compromised, it cannot be read by unauthorized parties.

When implementing QBD, businesses should also consider the security of the APIs being used. APIs are a common entry point for cyberattacks, so it is crucial to ensure that they are secure. One way to do this is to use API gateways, which can provide an additional layer of security by filtering traffic and blocking any requests that are deemed malicious.

Another important consideration when implementing QBD is privacy regulations. As businesses exchange data with other businesses, they must ensure that they are in compliance with all applicable privacy laws and regulations. This can be a complex task, as there are many regulations around the world, each with their own requirements. However, by implementing a robust consent management system, businesses can ensure that they are collecting and processing data in accordance with regulations.

To provide the highest level of security, QBD services use the most advanced security protocols available. For example, our QBD services use Sig 4 and Libsodium, which are state-of-the-art cryptographic libraries that provide robust security against a wide range of attacks. Additionally, we use one-way hashing to protect data, which means that data cannot be decrypted, even if it is compromised.

Finally, businesses should also consider the importance of data governance. Data governance involves the management of data, including data quality, data security, and compliance with regulations. By implementing strong data governance practices, businesses can ensure that their data is of high quality and is being used in compliance with regulations.

In summary, the security and privacy of data are critical considerations when implementing QBD systems and creating bridges with other businesses. Businesses must ensure that their own data is secure before engaging in QBD, and should implement strong access controls and encryption to protect their data. Additionally, businesses should consider the security of the APIs being used, comply with privacy regulations, and implement robust data governance practices. By following these best practices, businesses can ensure that their data is safe and secure, and can take full advantage of the benefits of QBD.

C. Establishing performance metrics

Establishing performance metrics is a critical aspect of implementing QBD for your business. Without clear metrics in place, it can be challenging to measure the effectiveness of your QBD implementation and track your progress toward achieving your business goals.

One essential aspect of establishing performance metrics is setting specific, measurable, achievable, relevant, and time-bound (SMART) goals. For instance, you may set a goal to increase revenue by 10% through the implementation of QBD within the next six months. This goal is specific in terms of revenue, measurable in terms of a 10% increase, achievable with the implementation of QBD, relevant to the business's overall goals, and time-bound within six months.

Another crucial aspect of establishing performance metrics is defining the key performance indicators (KPIs) that will help you measure your progress towards your goals. Some examples of KPIs for QBD implementation might include the number of successful Bridges established with external businesses, the number of Bi-link tokens created, the time and cost savings achieved through QBD implementation, and the number of new revenue opportunities generated through QBD.

To track and measure these KPIs, you will need to implement robust analytics and reporting tools. These tools will allow you to collect and analyze data related to your QBD implementation, visualize it in easy-to-understand dashboards, and make data-driven decisions to improve performance and achieve your goals.

In addition to setting SMART goals and defining KPIs, it is also essential to establish a performance management system that includes regular monitoring, analysis, and reporting of your QBD implementation's progress. This system should involve regular check-ins with stakeholders, including executives, project managers, and key team members, to review performance metrics and adjust strategies and tactics as needed.

As part of your performance management system, it's also essential to establish a culture of continuous improvement, where all team members are encouraged to share ideas and feedback to improve QBD implementation continually. This culture of continuous improvement can help your business stay agile and adaptable in a rapidly changing business environment, making it easier to achieve your goals and maintain a competitive advantage.

Finally, it's crucial to communicate your QBD performance metrics and progress towards your goals to all stakeholders, including employees, customers, partners, and investors. This communication helps to build trust and transparency and can lead to increased support for QBD implementation initiatives.

In summary, establishing performance metrics is a critical aspect of implementing QBD for your business. By setting SMART goals, defining KPIs, implementing robust analytics and reporting tools, establishing a performance management system, supporting a culture of continuous improvement, and communicating progress to stakeholders, you can achieve your QBD implementation goals and maintain a competitive advantage in your industry.

V. Managing and Optimizing Bridges

A. Monitoring and analyzing performance data

Monitoring and analyzing performance data is crucial for ensuring the success of bridging partnerships established using QBD. When it comes to monitoring and analyzing performance data, AWS offers several powerful tools that can be leveraged to achieve this goal.

One such tool is Amazon CloudWatch, a monitoring and management service that allows businesses to collect and track metrics, collect and monitor log files, and set alarms. With CloudWatch, businesses can easily monitor the performance of their bridging partnerships, tracking metrics such as latency, error rates, and more. This information can be used to identify potential issues and take corrective action to optimize performance.

In addition to CloudWatch, AWS also offers several other tools that can be used for monitoring and analyzing performance data. Amazon Kinesis, for example, is a platform for streaming data on AWS, allowing businesses to easily collect and analyze large amounts of data in real-time. With Kinesis, businesses can monitor the performance of their bridging partnerships in real-time, identifying potential issues and taking corrective action before they have a significant impact on performance.

Another powerful tool for monitoring and analyzing performance data is AWS X-Ray. This service allows businesses to trace requests made to their bridging partnerships, identifying the root cause of issues and visualizing the dependencies between different components of the system. With X-Ray, businesses can quickly identify bottlenecks and other performance issues, optimizing the performance of their bridging partnerships.

To further optimize the performance of bridging partnerships, businesses can use AWS Lambda, a serverless computing platform that allows businesses to run code without provisioning or managing servers. With Lambda, businesses can easily automate the monitoring and analysis of performance data, using event-driven computing to trigger analysis when specific performance metrics are met or exceeded.

In addition to these tools, AWS also offers a range of other services and solutions that can be used to monitor and analyze performance data, depending on the specific needs of the business. With these tools and solutions, businesses can optimize the performance of their bridging partnerships, ensuring that they are delivering maximum value to all parties involved.

Overall, monitoring and analyzing performance data is critical for the success of bridging partnerships established using QBD. With the tools and services provided by AWS, businesses can easily track and optimize the performance of their bridging partnerships, ensuring that they are delivering maximum value to all parties involved.

B. Addressing issues and resolving conflicts

As a QBD user, it is important to be aware that issues and conflicts can arise in the bridging process. This is a normal part of any business relationship, and it is essential to address these issues promptly and efficiently to maintain a healthy business partnership. Here are some helpful self-help techniques that can save you time and money when addressing issues and resolving conflicts:

1. Communicate openly and honestly: The first step in resolving conflicts is to communicate openly and honestly with your business partner. Be clear and concise about the issues you are facing and try to avoid any confrontational language. Instead, focus on finding common ground and working towards a mutually beneficial solution.
2. Use mediation services: If you are having trouble resolving conflicts with your business partner, consider using mediation services. This can be an effective way to resolve conflicts without resorting to legal action, and it can save you both time and money in the long run.
3. Be proactive: Don't wait for issues to arise before addressing them. Instead, be proactive in your approach to business partnerships. Regularly review your partnership agreement and address any potential issues before they become a problem.
4. Stay focused on your goals: When issues arise, it can be easy to get sidetracked and lose sight of your goals. Stay focused on your long-term goals and the benefits of the business partnership, and work towards finding a solution that benefits both parties.
5. Use QBD tools for monitoring and analysis: QBD provides tools for monitoring and analyzing performance data, which can help you identify issues before they become a problem. By monitoring your bridging performance data, you can stay on top of any potential issues and address them before they become a major problem.
6. Consider legal action as a last resort: Legal action should be a last resort when resolving conflicts with your business partner. Instead, try to find a mutually beneficial solution through open and honest communication and mediation services.

Addressing issues and resolving conflicts is an essential part of any business partnership. By using these self-help techniques and staying focused on your long-term goals, you can maintain a healthy business relationship with your partner and save yourself time and money in the process. Remember, communication is key, and by using QBD tools for monitoring and analysis, you can stay on top of any potential issues and address them before they become a major problem.

C. Adjusting bridging protocols and terms as needed

Adjusting bridging protocols and terms as needed is a critical part of the QBD process. As businesses grow and evolve, their needs and priorities change. Similarly, the priorities and needs of their partners may also change, and the existing bridging protocols may no longer be effective. In such cases, it is important to revisit the bridging protocols and terms and adjust them as needed to ensure that they continue to support the business goals of all parties involved.

One example of when it may be necessary to adjust bridging protocols and terms is when there is a change in the regulatory landscape. Laws and regulations governing data privacy and security are constantly evolving, and businesses must stay up to date to ensure compliance. If a new regulation is enacted that affects the data shared between two partners, it may be necessary to adjust the bridging protocols and terms to ensure compliance with the new regulation.

Another example is when there is a change in business strategy. Businesses may shift their priorities or focus on new areas, which may require a different approach to bridging protocols and terms. For example, if a business decides to expand into a new market, it may need to adjust the terms of its existing partnerships or establish new partnerships to support its expansion goals.

Additionally, changes in technology can also necessitate adjustments to bridging protocols and terms. New technologies may provide opportunities for more efficient and effective data sharing, but they may also require new protocols or terms to ensure data security and privacy. For example, the implementation of blockchain technology may provide new opportunities for secure and efficient data sharing, but it may also require new protocols and terms to ensure the integrity of the shared data.

To adjust bridging protocols and terms as needed, it is important to have a clear understanding of the existing protocols and terms and how they support the business goals of all parties involved. Businesses should regularly review their bridging protocols and terms to ensure that they are still effective and relevant. This review should include a thorough analysis of the data being shared, the security and privacy measures in place, and any regulatory requirements that may apply.

Once any necessary adjustments have been identified, businesses should work collaboratively with their partners to implement the changes. Communication is key in this process, and both parties should be involved in the decision-making process to ensure that the new protocols and terms are mutually beneficial.

One technique that can be helpful in adjusting bridging protocols and terms is to conduct a cost-benefit analysis. This analysis can help businesses identify the potential costs and benefits of the proposed changes and can provide valuable insights into the potential impact of the changes on the business and its partners.

Overall, adjusting bridging protocols and terms as needed is a critical component of the QBD process. By regularly reviewing and updating these protocols, businesses can ensure that their partnerships continue to support their evolving needs and priorities, while also maintaining compliance with regulatory requirements and ensuring the security and privacy of shared data.

VI. Conclusion

A. Recap of key points

In conclusion, the Bridging Your Business To Other Businesses methodology with QBD provides businesses with a powerful toolset for creating and managing partnerships with other businesses. The key points to remember are:

1. QBD provides businesses with a standardized method for creating and managing bi-directional link chains between multiple parties.
2. Businesses should identify potential Targets for partnerships based on complementary products, services, or customer bases.
3. Prioritizing potential Targets can be achieved by using a combination of internal data analysis, market research, and expert consulting.
4. When creating Bridges, businesses should determine value exchanges, negotiate terms and agreements, and establish bridging protocols that meet the needs of all parties.
5. To ensure the success of these partnerships, businesses should establish performance metrics, monitor and analyze performance data, and address issues and conflicts as they arise.
6. Adjusting bridging protocols and terms as needed is essential to ensure that the partnership remains profitable and beneficial to all parties involved.

By following these key points, businesses can take advantage of the opportunities that partnerships with other businesses present. With QBD, businesses can create and manage these partnerships more efficiently and effectively, ultimately resulting in increased revenue and growth.

B. Future directions and possibilities

As QBD becomes more widely adopted, businesses will have the opportunity to explore a wide range of possibilities for bridging their operations with other businesses. Here are a few exciting possibilities to consider:

1. Enhanced Supply Chain Management: QBD can be used to create

Bi-link chains between suppliers, manufacturers, distributors, and retailers. This can provide real-time visibility into the entire supply chain, allowing for more accurate demand forecasting, better inventory management, and reduced waste.

2. Innovative Customer Engagement: QBD can be used to create Bi-link chains between businesses and their customers, allowing for more personalized and interactive engagement. For example, a hotel could use QBD to create a Bi-link chain with local restaurants, museums, and other attractions, providing guests with personalized recommendations and exclusive discounts.
3. Data Sharing for Research and Development: QBD can be used to create Bi-link chains between businesses and research institutions, enabling secure and efficient data sharing. This can facilitate collaborative research and development projects, leading to breakthrough innovations and new products and services.
4. Integrated Healthcare Ecosystems: QBD can be used to create Bi-link chains between healthcare providers, insurance companies, and pharmaceutical companies. This can enable more efficient and effective healthcare delivery, improved patient outcomes, and reduced costs.
5. Smart Cities: QBD can be used to create Bi-link chains between city governments, businesses, and residents, enabling the creation of smart cities. This can lead to more efficient and sustainable urban environments, with improved transportation, energy management, and public services.

As businesses continue to explore the possibilities of QBD, new and innovative use cases will emerge, leading to a more interconnected and collaborative business ecosystem.

C. Call to action for businesses to start bridging

Are you ready to take your business to the next level and unlock its full potential? It's time to start bridging! By implementing QBD solutions and integrating with complementary businesses, you can expand your market reach, improve your products and services, and increase your revenue streams.

Think about the possibilities of collaborating with other businesses in your industry or even outside of it. Imagine the new innovations that can be created, the customer experiences that can be enhanced, and the efficiencies that can be gained. With QBD, the opportunities are endless.

Don't wait for your competitors to take the lead. Start bridging today and position your business as a leader in your industry. With QBD's easy-to-adopt APIs, OpenAPI documentation, AI seed data, and human control panels, you can quickly and securely integrate with other businesses.

Plus, with our highest security protocols and consent mechanism, you can ensure the protection of your sensitive data and maintain compliance with all privacy regulations. Our performance metrics and monitoring tools allow you to optimize your bridging strategy and adjust as needed.

So, take the first step towards bridging your business to other businesses and unlock the full potential of your company. Our team of experts is ready to guide you through the process and help you achieve success. Don't wait, start bridging today!

Developing Targets to Attract Business

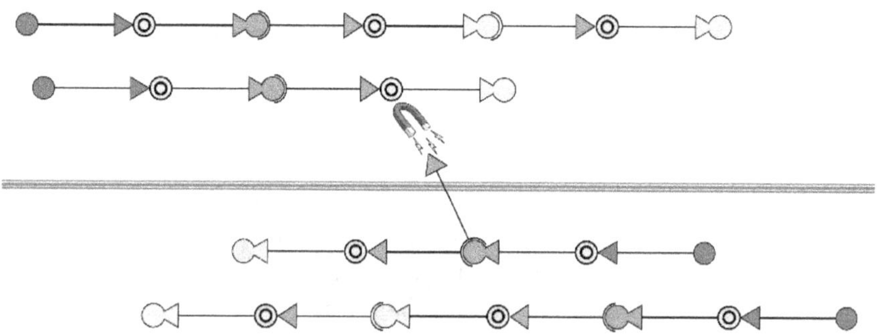

I. Introduction

A. Brief overview of the importance of attracting businesses to your own business

Attracting other businesses to your own is a critical component of achieving growth and success in today's highly competitive business landscape. By forming strategic partnerships with complementary businesses, you can leverage each other's strengths, gain access to new customers and markets, and create innovative solutions that benefit all parties involved.

B. Explanation of how QBD can facilitate the process of attracting businesses

Quantum Business Dynamics™ (QBD™) is a powerful tool that can help businesses attract other businesses in a more efficient and effective manner. By leveraging QBD, businesses can develop a clear understanding of their own unique strengths and weaknesses, as well as those of their potential partners.

This understanding can then be used to develop a targeted strategy for attracting businesses that are complementary to your own, and can provide the resources, expertise, and capabilities needed to help you achieve your business goals. QBD can also help businesses create a framework for establishing and managing these partnerships in a way that is mutually beneficial and sustainable over the long term.

Here are some specific ways that QBD can facilitate the process of attracting businesses:

1. Identify potential partners: By using QBD, businesses can identify potential partners that are a good fit based on a variety of factors, including their capabilities, expertise, and business model. This can help businesses narrow their focus and identify the most promising Targets for partnership.
2. Develop a clear value proposition: QBD can help businesses develop a clear value proposition that outlines the benefits of partnering with your business. This can help you stand out from the competition and demonstrate the unique value that your business can bring to a partnership.
3. Create a structured approach to partnership development: By using QBD to create a structured approach to partnership development, businesses can ensure that they are targeting the right partners, and that they are developing partnerships in a way that is mutually beneficial and sustainable over the long term.
4. Streamline partnership negotiations: QBD can also help businesses streamline the process of negotiating partnerships by providing a clear framework for establishing terms and conditions, and by identifying potential areas of conflict or disagreement in advance.

Overall, by leveraging QBD, businesses can improve their chances of attracting the right partners and creating successful partnerships that drive growth and success over the long term.

II. Identifying Potential Targets

A. Identifying complementary businesses through QBD analysis

One of the key benefits of QBD is the ability to identify complementary businesses that can help your business grow. Through QBD analysis, businesses can identify potential partners that can add value to their own products or services. By identifying complementary businesses, businesses can create partnerships that benefit both parties, increase market share, and enhance customer satisfaction.

To identify complementary businesses, QBD analysis can be performed to identify businesses with similar customer demographics, complementary product or service offerings, and similar business models. By leveraging QBD analysis, businesses can identify partners that can help them create unique and compelling value propositions for their customers.

B. Utilizing QBD tools and AI to gather data on potential Targets

QBD tools and AI can also be utilized to gather data on potential Targets. This can include data on their customer demographics, product and service offerings, business models, and market share. By utilizing QBD tools and AI, businesses can gather data on potential partners quickly and easily, saving time and resources.

QBD tools and AI can also help businesses identify potential partners that may not be obvious at first glance. By analyzing large sets of data, QBD tools and AI can identify trends and patterns that may not be immediately apparent to human analysts.

C. Conducting market research and analysis to identify potential Targets

Market research and analysis can also be utilized to identify potential Targets. This can include analyzing market trends, customer behavior, and competitor activity. By conducting market research and analysis, businesses can gain a deeper understanding of their target markets and identify potential partners that can help them capture more market share.

D. Applying QBD methodology to prioritize potential Targets

Once potential Targets have been identified, QBD methodology can be applied to prioritize them. This involves analyzing potential partners based on a number of factors, such as the strength of their customer base, the value of their product or service offerings, and the potential for long-term partnerships.

By applying QBD methodology to prioritize potential Targets, businesses can focus their resources on the partners that are most likely to add value to their business. This can help businesses create strong and sustainable partnerships that benefit both parties.

Overall, QBD can be a powerful tool for identifying potential Targets and attracting businesses to your own business. By utilizing QBD analysis, tools, and methodology, businesses can create partnerships that enhance their value proposition and drive growth.

III. Developing Strategies to Attract Targets

A. Determining value exchanges and potential benefits for potential Targets

When identifying potential Targets, it is important to determine what value exchanges and benefits your business can offer to them. Utilizing QBD methodology, analyze the potential target's business processes and identify areas where your business can provide unique value. Consider what resources, expertise, or solutions your business can offer to help solve their pain points or challenges. By offering a mutually beneficial value exchange, you can build a strong foundation for a successful partnership.

B. Building relationships with potential Targets through targeted marketing and communication

Once potential Targets have been identified, it's important to begin building a relationship with them. Utilize targeted marketing and communication strategies to introduce your business and the potential benefits of partnering with you. Consider attending industry events, networking with key players, and leveraging social media platforms to engage with potential Targets. By establishing a rapport and demonstrating your expertise, you can build trust and credibility with potential Targets.

C. Establishing shared goals and objectives with potential Targets

When developing strategies to attract potential Targets, it's important to establish shared goals and objectives. Identify areas where your business and the potential Target have common goals or objectives and work towards developing a shared vision for the partnership. By aligning your goals and objectives, you can create a strong foundation for a successful partnership that is focused on achieving mutual success.

D. Identifying areas where your business can provide unique value to potential Targets

To attract potential Targets, it's important to identify areas where your business can provide unique value. Utilize QBD methodology to identify areas where your business has a competitive advantage or unique expertise that can benefit potential Targets. By highlighting these areas of strength, you can demonstrate the value that your business brings to the partnership and differentiate yourself from competitors.

IV. Developing Potential Targets

By following these strategies, businesses can effectively identify potential Targets, develop strong relationships, and establish successful partnerships that benefit all parties involved. QBD methodology provides businesses with the tools and frameworks to successfully navigate the process of attracting and developing relationships with potential Targets.

A. Building a comprehensive understanding of potential target's needs and goals: In order to effectively negotiate terms and agreements with potential Targets, it is crucial to have a deep understanding of their needs, goals, and priorities. This involves conducting thorough research on the potential target's industry, market position, and business objectives. By gathering information about their values, strengths, and challenges, you can tailor your offers and proposals to better align with their needs and goals. This also involves understanding their decision-making process and identifying key decision-makers within their organization.

B. Crafting offers that align with both your business's goals and the potential target's goals: Once you have a comprehensive understanding of the potential target's needs and goals, you can craft offers and proposals that align with both your business's goals and the potential target's goals. This involves identifying areas of mutual benefit and developing creative solutions that address shared challenges. It is important to be flexible in your negotiations and open to compromise to reach mutually beneficial agreements. By demonstrating a willingness to collaborate and find common ground, you can build trust and strengthen the relationship between your businesses.

C. Establishing mutually beneficial terms and conditions for partnerships: The terms and conditions of a partnership agreement should reflect the mutual benefits and goals of both businesses. This involves negotiating specific terms such as revenue sharing, resource allocation, intellectual property rights, and other legal and financial considerations. By establishing clear and mutually beneficial terms, you can reduce the risk of misunderstandings or disagreements down the road. It is important to communicate openly and transparently throughout the negotiation process in order to build a strong foundation for the partnership.

D. Utilizing QBD tools and AI to ensure fair and equitable agreements: QBD tools and AI can be incredibly valuable in negotiating fair and equitable agreements with potential Targets. These tools can help you gather and analyze data, identify areas of mutual benefit, and simulate various scenarios to find the best possible outcomes for both businesses. By using QBD tools and AI, you can ensure that your negotiations are data-driven, objective, and grounded in evidence. Additionally, QBD tools and AI can help you identify potential risks and develop contingency plans to mitigate them.

Negotiating terms and agreements with potential Targets is a crucial aspect of bridging your business to other businesses. By building a comprehensive understanding of the potential target's needs and goals, crafting offers that align with both your business's goals and the potential target's goals, establishing mutually beneficial terms and conditions, and utilizing QBD tools and AI, you can negotiate fair and equitable agreements that strengthen partnerships and drive growth for both businesses.

V. Laying Out the Standards

Good fences make for good neighbors. There is always wisdom in thinking through the guidelines you will want your neighbors to follow along with you and your organization.

A. Developing a set of standard protocols for initiating and maintaining partnerships.

When establishing bridging protocols, it is important to develop a set of standard protocols for initiating and maintaining partnerships. This can include protocols for communication, data sharing, and collaboration. By developing standard protocols, businesses can ensure that their partnerships are consistent and reliable, reducing the risk of miscommunication or misunderstandings. These protocols can also help to establish trust between the parties involved, as they provide a clear framework for how the partnership will operate.

B. Establishing clear guidelines for data sharing and management

One of the most critical components of bridging protocols is establishing clear guidelines for data sharing and management. This is particularly important when it comes to sensitive or confidential data. It is essential that all parties involved understand how data will be shared and managed, including who will have access to it, how it will be stored, and how it will be protected. Clear guidelines can help to prevent data breaches and ensure that all parties are in compliance with relevant laws and regulations.

C. Ensuring that all parties involved are in compliance with relevant laws and regulations

When establishing bridging protocols, it is essential to ensure that all parties involved are compliant with relevant laws and regulations. This can include data protection regulations, intellectual property laws, and other relevant regulations. Failure to comply with these regulations can result in legal consequences, which can be costly for businesses. By ensuring that all parties are in compliance, businesses can minimize risk and establish a solid foundation for their partnerships.

D. Utilizing QBD tools and AI to facilitate seamless integration between businesses

Finally, QBD tools and AI can be invaluable in facilitating seamless integration between businesses. These tools can help to automate the process of data sharing and management, reducing the risk of errors or misunderstandings. They can also provide insights and analytics that can help businesses to identify areas for improvement and optimize their partnerships for maximum value. By utilizing these tools, businesses can establish efficient, effective, and profitable partnerships that are built to last.

VI. Conclusion

A. Recap of Key Points: Throughout this guide, we have discussed the importance of attracting businesses to your own business and how QBD can facilitate this process. We have explored how to identify potential Targets through QBD analysis, market research, and prioritization. We have also discussed how to negotiate terms and agreements and establish bridging protocols to ensure seamless integration between businesses.

B. Encouragement for Businesses to Use QBD Methodology: We strongly encourage businesses to use QBD methodology to develop and attract new partnerships. By leveraging QBD tools and AI, businesses can streamline the process of identifying, negotiating, and establishing partnerships, leading to faster growth and increased success.

C. Discussion of Potential Benefits: By implementing QBD methodology, businesses can tap into new revenue streams and expand their market reach. They can also gain valuable insights and knowledge from their partners, leading to increased innovation and competitiveness. QBD also provides a secure and compliant way to share data, ensuring that businesses can collaborate with confidence and ease.

In conclusion, QBD offers businesses an effective and efficient way to develop and attract new partnerships. By leveraging QBD tools and AI, businesses can streamline their processes and accelerate growth. We hope that this guide has been helpful and informative, and we encourage businesses to explore the many benefits of implementing QBD methodology.

How to maximize verification

In the world of business, it's important to have reliable data that you can trust. That's why verification is so critical in Bi-link chains. By verifying conclusions from the accuracy, reliability, and trustworthiness of data, protocols, and algorithms in a Bi-link chain, businesses can ensure that they are making informed decisions based on factual information.

The flexibility and power of verification in QBD cannot be overstated. It has the potential to drive growth in specific areas of a business by providing confidence in the data and facilitating the flow of money throughout the Bi-link Fabric.

To understand verification, it's important to grasp its purpose in a Bi-link chain. Verification is the process of confirming the accuracy, reliability, and trustworthiness of data, protocols, and algorithms in a Bi-link chain. Its purpose is to satisfy questions that can be served by gaining such confirmation, such as authenticity or validity of facts contained within the Bi-link chain. Verification is used to confirm agreements between two other humans regarding things such as the identity of parties involved in a Bi-link chain, to verify the authenticity of transactions, or to ensure that documents and other data are genuine and unaltered.

There are different types of verification, and they each have their own unique uses. For example, a bank HR professional may need to verify that an employee has been duly certified in something to satisfy regulatory requirements. On the other hand, an educational facility may want to pay for verification to bolster the reputation of their institution. The profits from verification can flow in many different directions, depending on the needs of the parties involved in the Bi-link chain.

The benefits of verification are numerous, and all parties involved in a Bi-link chain can benefit from it. The Authority that minted the Target can make their certification more valuable for relying parties such as banks and their employees. The validatory of that certification can bolster the reputation of their institution. The Owner, the student that took the course and passed the test, can keep their job. The Verifier, the HR professional or their employer, can stay compliant with regulations.

Verification is not only a powerful tool for ensuring accuracy and reliability in Bi-link chains, but it can also be a key driver of growth for businesses. By leveraging the flexibility of verification in QBD, businesses can use it to encourage growth in specific areas and direct the flow of money in their Bi-link chains.

One way that verification can drive growth is by increasing the value of a business's products or services. By verifying the authenticity and quality of their offerings, businesses can establish trust with their customers and differentiate themselves from competitors. For example, a clothing manufacturer can use verification to prove that their garments are made from sustainable materials and have been ethically produced, which can increase their appeal to environmentally conscious consumers.

In addition to enhancing the value of their products or services, verification can also be used to streamline processes and reduce costs. By ensuring that all parties in a Bi-link chain are operating on the same page, businesses can minimize errors and reduce the need for manual checks and redundancies. This cannot only improve efficiency but also free up resources to be allocated towards other growth areas.

Furthermore, the flow of money in a Bi-link chain can be directed through the implementation of verification. Businesses can leverage verification to incentivize certain behaviors or outcomes, such as offering discounts to customers who provide verified referrals or using verification to ensure that suppliers are meeting certain quality standards. By adjusting the parameters of verification in their Bi-link chains, businesses can align the interests of all parties involved and achieve their growth objectives.

To illustrate the power of verification in driving growth, consider the example of a small business that produces artisanal goods. By using verification to ensure the authenticity and quality of their products, they can establish a loyal customer base and differentiate themselves from competitors. They can also use verification to incentivize customers to provide verified referrals, which can drive growth in their sales. By directing the flow of money in their Bi-link chain towards their growth objectives, this small business can achieve sustainable growth and success.

As Authorities, it is important to consider certain factors when establishing Targets for verification within a Bi-link chain. One key consideration is to ensure that the verification process aligns with the goals and needs of your business. This means identifying areas where verification can drive growth and implementing it in a targeted and strategic way.

However, there are also potential pitfalls to be aware of. For example, over-reliance on verification can lead to increased costs and complexity, as well as potential concerns around privacy and data security. It is important to balance the benefits of verification with these potential drawbacks and ensure that it is being implemented in a responsible and effective manner.

To manage verification in a Bi-link chain, there are several tips that can be helpful. One important tip is to establish clear guidelines for data sharing and management, to ensure that all parties involved are in compliance with relevant laws and regulations. Additionally, utilizing QBD tools and AI can help to facilitate seamless integration between businesses and ensure fair and equitable agreements.

This leads us to a keener appreciation that verification is a powerful tool for driving growth and maximizing the benefits of Bi-link chains in QBD. However, it is important for Authorities to carefully consider their Targets and manage the process in a responsible and effective manner. By doing so, businesses can unlock new opportunities for growth and success.

Conclusion

As we approach the conclusion of this transformative methodology book, it is crucial to pause and reflect on the journey we have embarked upon together. We have explored the various aspects of QBD, delved into its implementation, and examined real-world examples to illustrate its potential impact on businesses. In this final section, we will recap the key points and provide a broader perspective for business executives to appreciate the value of QBD in their organizations.

I. Recap of Key Points

A. Summary of the main topics covered in the methodology

Throughout the book, we have covered a wide range of topics related to QBD. We began by discussing the three most important tasks for businesses: optimizing operations, mitigating risks, and unlocking new opportunities. We then explored the four biggest risks, including regulatory compliance, economic uncertainty, and tech and security threats, as well as the four biggest missed opportunities, such as missed innovation and growth, digital transformation, and operational cost savings.

We continued by introducing the four particles of QBD - Sources, Targets, Bridges, and Binary Links - and provided a comprehensive understanding of their roles in the methodology. In the implementation section, we discussed the benefits and challenges of QBD, best practices for implementation, and real-world examples of successful QBD adoption.

B. Explanation of the importance of QBD and its potential benefits for businesses

QBD is an essential methodology for modern businesses aiming to thrive in an increasingly complex and competitive landscape. It offers a unique approach to optimizing operations, mitigating risks, and unlocking new opportunities by leveraging the power of data, trust, and security.

The importance of QBD for businesses lies in its ability to enhance trust and security through its innovative approach to data management and exchange. By implementing QBD, organizations can:

1. Streamline operations: QBD enables businesses to optimize their processes by providing real-time monitoring, data validation, and access control mechanisms, leading to increased efficiency and productivity.
2. Mitigate risks: QBD helps organizations address various risks such as regulatory compliance, economic uncertainty, and tech and security threats by providing a secure and transparent environment for data management.
3. Unlock new opportunities: QBD can facilitate innovation and growth by enabling new business models based on data sharing and collaboration among multiple entities. Moreover, it can aid digital transformation efforts and help organizations achieve operational cost savings through improved data management.
4. Enhance interoperability: QBD's modular architecture and flexible workflows allow for seamless integration with existing systems and technologies, enabling businesses to maximize their investments in legacy infrastructure.

In conclusion, QBD offers a comprehensive methodology that can significantly benefit businesses by improving trust, security, and efficiency in their operations. By reflecting on the key points and insights gained throughout this book, business executives can better appreciate the potential impact of QBD on their organizations and make informed decisions on its implementation. As we wrap up, we encourage you to explore QBD further and consider the transformative potential it holds for your organization in today's rapidly evolving business landscape.

Final Thoughts

Let us take a moment to envision the future potential of this powerful methodology and how it can continue to revolutionize the way businesses operate. With its unique approach to data management, trust-building, and security enhancement, QBD promises to be a game-changer for organizations across industries.

The future potential of QBD lies in its ability to continuously adapt and evolve to meet the ever-changing needs of businesses. As technology advances and the business landscape becomes more complex, QBD can help organizations stay ahead of the curve by providing innovative solutions for data management, collaboration, and decision-making. By leveraging the power of QBD, businesses can achieve unparalleled levels of efficiency, trust, and security, setting themselves apart from competitors and paving the way for long-term success.

Now is the time for businesses to seize the opportunity and begin implementing QBD within their organizations. By doing so, they can unlock the immense potential of this methodology and set themselves on a path towards growth, innovation, and success. As more businesses adopt QBD, they can bridge with other organizations, undergirding a collaborative ecosystem that will drive the future of business in the digital age.

We encourage businesses to continue exploring and experimenting with QBD to unlock its full potential. As with any groundbreaking methodology, there will always be new discoveries, techniques, and possibilities to uncover. By staying open to learning and pushing the boundaries of what QBD can achieve, businesses can ensure they remain at the forefront of their industry.

In conclusion, QBD offers a transformative methodology that can revolutionize the way businesses operate in the modern world. Its future potential is vast, and it is up to organizations to seize this opportunity and begin implementing QBD within their operations. By embracing this methodology and bridging with other businesses, organizations can unlock untold possibilities for growth, innovation, and success. We urge you to take the first step on this exciting journey, and we look forward to witnessing the incredible impact QBD will have on the future of your business.

SECTION III – TEACHING QBD™ TO YOUR TEAM

A hands-on guide to rapid deployment!

What to expect

Brief overview

Welcome to Quantum Business Dynamics™ (QBD™), where you're about to embark on an exciting journey of transforming your business ecosystem. This section will guide you through the QBD implementation process, providing insights into benefits, potential risks, and practical industry-specific examples.

- QBD Methodology: Expect a clear, concise explanation of the QBD methodology. You'll learn how key components - Sources, Targets, Bridges, and Binary Links - work together, contributing to greater trust, transparency, and security in your business.
- Benefits and Potential Risks: Understand the benefits of QBD implementation and the potential challenges involved. This section discusses potential risks, helping you make an informed decision about QBD's suitability for your business.
- Industry-Specific Examples: Learn how QBD can tackle challenges in various sectors like agriculture and scientific services, providing more transparent and secure solutions.
- Preparation and Planning: Discover the necessary steps for QBD implementation, including stakeholder identification, resource allocation,

and the creation of a detailed implementation plan.

- Hurdles and Obstacles: Overcome common challenges like resistance to change and technical difficulties, ensuring smooth implementation of QBD.
- Training and Support: Get insights into providing effective training and support to your team, ensuring successful adoption of QBD methodology.
- Monitoring and Evaluation: Understand the importance of monitoring and evaluating QBD implementation success. Learn to identify areas for improvement and refine your approach for desired outcomes.

How QBD Can Benefit Businesses

Quantum Business Dynamics™ (QBD™) is a revolutionary way to manage business agreements. It bolsters security, trust, and transparency, minimizing risks and enhancing efficiency. This section explores the benefits of QBD across different sectors with successful implementation examples.

- Security: QBD provides robust security, critical for sectors like food services and insurance. QBD's ability to secure and verify agreements at every step mitigates the risk of fraudulent activities.

- Trust: QBD leverages a decentralized system to ensure transparency and verification of agreements. This feature builds trust, vital in industries like finance and healthcare, where stakes are high.

- Efficiency: QBD promotes efficiency in managing complex agreements. By automating processes using QBD, sectors like finance and healthcare can reduce costs and improve customer experiences.

QBD is indeed a potent tool that establishes trust and transparency in business relationships, improves security, and reduces fraud risk. As blockchain technology's adoption continues to grow, QBD is set to become an essential part of modern business operations.

Understanding QBD

If you are a teacher and skipped your way to this section, you will find this next section beneficial. However, if you have read from the beginning, feel free to turn ahead a few pages to "Challenges of Implementing QBD."

QBD is an innovative methodology constructed to enhance reliability and security in business transactions through the use of digital particles known as Sources, Targets, Bridges, and Binary Links. Rooted in the principles of quantum mechanics - which have radically transformed fields like computing and cryptography - QBD applies these principles to business transactions, fostering secure, unchangeable, and verifiable digital agreements.

Within the QBD framework, four types of digital particles - Sources, Targets, Bridges, and Binary Links - each hold unique properties that collectively enhance security and transparency.

Sources initiate QBD transactions, embodying an actionable validation, the veracity and history of which can later be relied upon as one half of an agreement.

Targets are the end goal of a QBD transaction. They symbolize the finalized agreement between the parties, encapsulating all necessary details regarding the transaction's outcome.

Bridges are the links connecting Sources and Targets, signifying brokered validation of an agreement between the parties involved. This structure provides a secure and transparent record of the transaction, embodying the original agreement between two parties, complete with the necessary details about the terms of the agreement, the parties involved, and other relevant specifics.

Binary Links are digital tokens born when Bridges are formed. They represent the value of the unique relationship between transaction parties, providing a possessable version of that value that is a secure and unchangeable agreement record.

Utilizing these four particle types, QBD delivers increased security, transparency, and trust in digital transactions to businesses.

Why QBD is Crucial for Businesses

QBD represents not only an innovative approach but a vital tool for modern businesses navigating an intricate digital landscape where trust, transparency, and security are key. QBD is especially useful in preventing fraudulent activities, as it creates an immutable and transparent transaction record, paramount in industries like finance and insurance where fraud can be prevalent.

QBD's applications span beyond fraud prevention to industries like supply chain management, logistics, and energy production, where it can create unalterable records of maintenance and safety inspections. It streamlines processes, reduces administrative burden, and facilitates new business model creation. As industries increasingly move toward digitalization, QBD's importance will only grow.

Benefits of QBD

Implementing QBD presents a myriad of benefits for businesses, such as enhanced transparency, increased security, and efficiency gains. More than that, it offers unique opportunities for businesses to innovate and set themselves apart in a competitive market.

Transparency, provided by QBD, can revolutionize industries like construction and energy, where projects often involve multiple stakeholders. Using Bridges and Binary Links to track project progress allows businesses to gain a real-time view of the entire supply chain, potentially identifying and addressing issues before they escalate.

Security-wise, QBD creates an unchangeable record of transactions and agreements, ensuring data protection from tampering or corruption, a particularly critical feature in sectors like finance and healthcare where data breaches can have severe implications.

Efficiency-wise, QBD streamlines agreement creation and verification, cutting administrative overhead and saving businesses time and money, a significant boon for SMEs with limited resources.

Additionally, QBD can increase trust and security, as exemplified by a Dark Fiber provider. In the capacity planning case, complex Bi-link chains effectively solve capacity planning issues while fostering trust and security. The chain begins with a Source (the provider), creates Bridges to Targets (potential customers), and follows customers' usage data through Binary Links. These Links provide information for capacity planning tools that predict future capacity needs. The efficient, transparent, and trust-building use of complex Bi-link chains can serve as a model for other industries, demonstrating the power and potential of QBD.

Challenges of Implementing QBD

QBD is a powerful tool that can help businesses improve their processes, increase transparency, and build trust with their customers. However, like any new methodology or approach, businesses may face challenges during the implementation process. In this section, we will explore some of the potential challenges businesses may encounter when implementing QBD and provide guidance on how to overcome them.

1. Resistance to change

One of the most common challenges businesses face when implementing QBD is resistance to change. Many employees may be resistant to adopting a new approach or methodology, especially if they are used to doing things a certain way. To overcome this challenge, it's important to communicate the benefits of QBD clearly to employees and provide them with training and support to help them adjust to the new approach.

2. Lack of understanding

Another challenge businesses may face is a lack of understanding about how QBD works and how it can benefit their organization. To overcome this challenge, it's important to provide clear and concise information about the methodology, including how it works and how it can be applied to specific business processes.

3. Integration with existing systems

Integrating QBD with existing systems and processes can be another challenge businesses may face. It's important to identify potential integration issues early on in the implementation process and work to address them before they become major obstacles. This may require additional training or the use of third-party tools and services to facilitate integration.

4. Data quality and management

Effective use of QBD requires high-quality data, and businesses may face challenges with data quality and management. It's important to establish clear data governance policies and procedures and invest in tools and resources to ensure data quality and accuracy.

5. Limited resources

Limited resources, such as time and budget, can also be a challenge when implementing QBD. It's important to prioritize implementation efforts and focus on high-value processes and use cases to maximize the impact of limited resources.

6. Resistance from external partners

External partners, such as suppliers or customers, may also be resistant to adopting QBD, which can create additional challenges for businesses. To overcome this challenge, it's important to communicate the benefits of QBD to external partners and work collaboratively to integrate the methodology into shared processes and systems.

7. Legal and regulatory compliance

Finally, legal and regulatory compliance can be a challenge when implementing QBD, especially in highly regulated industries such as healthcare or finance. It's important to work with legal and regulatory experts to ensure compliance with applicable laws and regulations.

Discussion of ways to overcome these challenges

We've listed above 7 challenges you may face when implementing a new methodology like QBD. However, with careful planning and execution, businesses can overcome these challenges and successfully implement QBD. In this section, we will discuss some ways to overcome these challenges.

1. Overcoming Resistance to Change

To address resistance to change, it is crucial to adopt a comprehensive change management strategy such as the ADKAR model developed by Prosci. This model focuses on five key elements: Awareness, Desire, Knowledge, Ability, and Reinforcement. By addressing each element, you can help employees transition more smoothly to QBD.

First, create Awareness by explaining the need for change and the benefits QBD brings to the organization. Next, engender Desire by emphasizing how QBD can improve individual performance and job satisfaction. Third, provide Knowledge by offering targeted training and resources to help employees understand QBD concepts and practices. Fourth, develop Ability by providing ongoing support and coaching to help employees apply QBD effectively. Finally, implement Reinforcement by recognizing and rewarding employees who embrace the change and contribute positively to the QBD implementation.

2. Addressing Lack of Understanding

To overcome a lack of understanding, use storytelling techniques to illustrate the value of QBD. Share real-life success stories from other organizations and case studies that showcase the tangible benefits of QBD. Additionally, create visual aids such as infographics, flowcharts, or videos to simplify complex concepts and make them more accessible to employees.

Furthermore, provide opportunities for hands-on learning through workshops, simulations, or pilot projects, allowing employees to experience the benefits of QBD firsthand. This experiential learning can lead to greater understanding and buy-in.

3. Facilitating Integration with Existing Systems

To streamline integration with existing systems, employ a phased approach, starting with smaller, less complex processes before tackling more complicated ones. This allows employees to build confidence and expertise as they learn to apply QBD in real-world scenarios.

Also, consider using Application Programming Interfaces (APIs) or middleware to bridge the gap between QBD and legacy systems, ensuring seamless data exchange and minimal disruption to existing workflows.

4. Ensuring Data Quality and Management

To tackle data quality and management challenges, adopt a "data-centric" mindset by emphasizing the importance of data integrity and accuracy throughout the organization. Implement strong data validation processes, automated data cleansing tools, and periodic data audits to ensure data remains accurate, complete, and up-to-date.

Additionally, undergird a data-driven culture by encouraging employees to use data for decision-making and providing them with the necessary tools, training, and resources to make data-driven decisions.

5. Maximizing Limited Resources

When faced with limited resources, adopt the Agile methodology to prioritize tasks and focus on high-impact initiatives. Break projects into smaller, manageable sprints, allowing for rapid iteration and continuous improvement.

Moreover, consider partnering with external consultants or leveraging existing resources within the organization, such as QBD champions or internal subject matter experts, to maximize the effectiveness of limited resources.

6. Overcoming Resistance from External Partners

To address resistance from external partners, invest in relationship-building and trust. Establish regular communication channels, such as status meetings, webinars, or workshops, to keep partners informed about QBD's benefits and progress.

Additionally, consider developing joint pilot projects or case studies with partners to showcase the value of QBD in a collaborative environment. These shared success stories can help to build trust and demonstrate the advantages of QBD adoption.

7. Navigating Legal and Regulatory Compliance

To ensure legal and regulatory compliance, adopt a proactive approach by staying abreast of changes in relevant laws and regulations. Establish a cross-functional compliance team that includes legal, regulatory, and QBD experts to conduct regular assessments and identify potential risks.

Moreover, invest in compliance training for employees, emphasizing the importance of adherence to legal and regulatory requirements in the context of QBD. By encouraging a culture of compliance, you can minimize risks and ensure the successful implementation of QBD.

Best Practices for Implementing QBD

Yes, Quantum Business Dynamics™ (QBD™) is great. However, implementing QBD is not without its challenges. To successfully implement QBD, businesses must follow a set of best practices to ensure that the process goes smoothly and yields the desired results.

The following is an overview of the best practices for businesses to follow when implementing QBD:

1. Start with a clear strategy: Before embarking on any new initiative, it is important to have a clear strategy in place. This includes identifying the business goals and objectives that QBD will help to achieve, as well as the specific processes and workflows that will be impacted by the implementation. This strategy should be communicated clearly to all stakeholders and serve as the foundation for the implementation process.
2. Involve all stakeholders: The success of QBD implementation depends on the involvement of all stakeholders, including employees, customers, partners, and suppliers. These stakeholders should be included in the planning process, as well as in the testing and deployment of the QBD solution. This will help to ensure that everyone is on board with the new approach and that the implementation is successful.
3. Ensure adequate training: QBD is a new approach that may require employees to learn new skills and processes. It is important to provide adequate training to all employees who will be impacted by the implementation, including those who will be responsible for managing the QBD solution on a day-to-day basis. This will help to ensure that everyone is prepared to use the solution effectively and that the implementation is successful.
4. Develop a testing and deployment plan: In order to ensure that the QBD implementation is successful, it is important to develop a

testing and deployment plan. This plan should include testing the solution in a controlled environment before deploying it to production, as well as a rollout plan that includes training, support, and feedback mechanisms. By following a well-developed plan, businesses can ensure that the QBD implementation is successful and yields the desired results.

5. Ensure ongoing support and maintenance: Once the QBD solution is deployed, it is important to ensure ongoing support and maintenance. This includes providing helpdesk support to users, as well as ongoing maintenance and upgrades to the QBD solution. By providing ongoing support and maintenance, businesses can ensure that the QBD solution continues to deliver value over the long term.
6. Measure success: Finally, it is important to measure the success of the QBD implementation. This includes defining key performance indicators (KPIs) that will be used to measure the success of the implementation, as well as regular reporting on these KPIs. By measuring success, businesses can identify areas for improvement and ensure that the QBD solution is delivering the desired results.

Implementing QBD can be a complex process that requires careful planning, stakeholder involvement, adequate training, and ongoing support and maintenance. By following the best practices outlined above, businesses can ensure that the implementation process goes smoothly and yields the desired results.

Explanation of how to ensure successful adoption of QBD

Successful adoption of QBD requires a well-planned and well-executed implementation process. In this section, we will discuss some key steps businesses can take to ensure successful adoption of QBD.

1. Define clear goals and objectives: Before embarking on the implementation process, it is important to define clear goals and objectives for the organization. These goals and objectives should be aligned with the overall strategy of the organization and should be communicated clearly to all stakeholders.
2. Engage stakeholders: It is important to engage all stakeholders in the implementation process, including employees, customers, partners, and vendors. By involving all stakeholders, businesses can ensure that everyone is aligned and committed to the implementation process.
3. Establish a governance structure: Establishing a governance structure is critical to the success of the implementation process. The

governance structure should define roles and responsibilities, decision-making processes, and communication protocols.

4. Develop a project plan: Developing a project plan is essential to the success of the implementation process. The project plan should include timelines, milestones, and deliverables.
5. Provide adequate training and support: Adequate training and support are essential to the successful adoption of QBD. This includes training on the methodology, tools, and processes, as well as ongoing support to ensure that employees are able to use QBD effectively.
6. Monitor progress: It is important to monitor progress throughout the implementation process to ensure that the project is on track and to identify any issues or challenges that need to be addressed.
7. Celebrate successes: Celebrating successes along the way is an important part of the implementation process. It helps to build momentum and keep stakeholders engaged and motivated.
8. Continuously improve: Continuous improvement is essential to the long-term success of QBD. Businesses should regularly review and evaluate the implementation process and make adjustments as necessary to ensure that QBD continues to meet the needs of the organization.

Successful adoption of QBD requires a well-planned and well-executed implementation process. By following these best practices, businesses can ensure that they are able to successfully adopt QBD and realize the benefits of this powerful methodology.

Real-world Examples of QBD in Action

Small to medium-sized businesses have increasingly turned to Quantum Business Dynamics™ (QBD™) to improve their operations and increase their competitiveness. One successful example is a company that builds small, controlled technology platforms for 150 to 250 small to medium-sized competitive companies that form a cooperative association. This association helps to improve compliance and share common aspects of sales channels while remaining competitive.

Their first implementation is a Cannabis farming community, but they plan to branch out into transportation and later insurance companies. QBD is fundamental to their technology platforms, and thus an organic part of all those businesses across industries. This provides a great opportunity for the three different networks to interconnect in new and meaningful ways through standard interfaces and agreements afforded to them by QBD.

By using QBD, this company can streamline their operations and improve their compliance with industry regulations. They can also improve their communication with their partners and customers through the secure and transparent digital platform that QBD provides. This improves trust and accountability, leading to stronger business relationships and greater efficiency.

Another successful implementation of QBD can be seen in the healthcare industry. Many healthcare providers have turned to QBD to improve patient outcomes and reduce costs. By using QBD to create immutable records of patient data and medical histories, healthcare providers can more easily share this information with other providers and improve the accuracy of diagnoses and treatments.

In addition, QBD can be used to streamline the billing and insurance claims process, reducing the administrative burden on healthcare providers and improving the experience for patients. This leads to better patient outcomes and a more efficient healthcare system overall.

Finally, QBD has also been successfully implemented in the financial services industry. Many financial institutions have used QBD to improve their security and compliance with regulations, while also improving their operational efficiency. By using QBD to create immutable records of financial transactions, institutions can more easily track and monitor their activities, reducing the risk of fraud and improving their compliance with regulations.

Overall, the implementation of QBD has provided many businesses with the tools they need to streamline their operations, improve their compliance with regulations, and increase their efficiency and competitiveness. Through standard interfaces and agreements, QBD provides a secure and transparent platform for businesses to communicate with their partners and customers, leading to stronger relationships and improved outcomes.

Explanation of the benefits they have experienced and how they overcame challenges

The company that builds small, controlled technology platforms for small to medium-sized competitive companies found QBD to be fundamental to their technology platforms. By using QBD, they were able to improve compliance and share common aspects of sales channels while remaining competitive.

One of the benefits they experienced was increased transparency in their supply chain, which helped build trust with their customers. They were also able to streamline their internal processes, reducing the administrative burden and costs. In addition, QBD allowed them to easily and securely share information and data with their partners, which helped improve collaboration and efficiency.

However, they did face some challenges when implementing QBD. One of the main challenges was ensuring that all partners and stakeholders were on board and willing to use the QBD platform. This required clear communication and education about the benefits of using QBD and how it would impact their businesses.

Another challenge was ensuring the security of the information and data being shared on the platform. They overcame this challenge by implementing robust security measures, including encryption and access controls, and by regularly monitoring the platform for any suspicious activity.

Overall, the company found that using QBD allowed them to improve their business processes and build stronger relationships with their partners and customers.

Bringing it all together

You've just read an extensive chapter on Quantum Business Dynamics™ (QBD™) and its effective implementation in businesses. This chapter outlined the notable benefits of QBD, from bolstering trust, transparency, and security in operational procedures, to increasing the efficiency of business processes.

We dove into the challenges that businesses could face while implementing QBD and offered strategies for overcoming those hurdles. The adoption of QBD might not always be straightforward, but through effective communication and thorough training, businesses can begin to create a culture of trust and transparency, integrating QBD into the company's overall strategy.

Practical examples were given throughout this chapter, illustrating how QBD has been successfully implemented in diverse industries. A highlighted case was that of a small technology platform company serving small and medium-sized businesses in the cannabis, transportation, and insurance industries. Their QBD adoption led to notable benefits such as improved compliance, heightened efficiency, and the emergence of new opportunities to interconnect networks.

In conclusion, this chapter stressed the considerable potential of QBD for businesses and provided practical guidance for successful implementation, presenting a clear roadmap for businesses ready to embark on their QBD journey.

Onboarding Your Team

In this section, we are going to provide you with a syllabus and scripts that you can follow to effectively teach your team what they need to know in order to successfully implement QBD in your business. As you read through it, our aim is for you to immediately see the value of QBD to your business by visualizing your teams learning it as they start to build out their own Bi-link Fabric.

Introduction

Let us peek in on an onboarding session for Quantum Business Dynamics™ (QBD™). This training session will provide you with an overview of QBD and how it can help you improve your business processes. By the end of this session, you will have a clear understanding of what QBD is, how it works, and how it can be implemented in your organization.

Session 1: Introduction to QBD

In this session, we will cover the basics of QBD and its benefits.

- Start by introducing yourself and the session.
- Explain what QBD is and how it works.

Alright folks, it's time to learn about something truly exciting today! Are you ready to dive into the world of QBD? I hope you are because we're going to have some fun.

So, what is QBD? QBD stands for "Quantum Business Dynamics™." It's a new and innovative business methodology that's all about bringing people and systems together to work more efficiently and effectively. Think of it as a big puzzle, with each piece representing a different aspect of your business. QBD is all about fitting those pieces together in a way that makes sense and creates a seamless and cohesive whole.

But how does it work? Well, let me ask you a question: have you ever played Jenga? You know...that game where you stack wooden blocks on top of each other until someone knocks it all down? QBD is kind of like that, but instead of wooden blocks, we're talking about different parts of your business.

Imagine that each block is a different department, process, or system within your organization. Each block is important on its own, but when you stack them all together in the right way, something magical happens.

Your business becomes more efficient, more effective, and more profitable.

But the real magic of QBD is in the Bi-links. Bi-links are like the glue that holds everything together. They're the connections between different parts of your business that allow them to work together seamlessly. And when you have a lot of Bi-links, your business starts to look like a spider web, with each strand connected to the next.

So, let's say you have a customer who wants to buy something from you. In a traditional business model, that customer would have to go through a bunch of different departments and processes before their order is fulfilled. But with QBD, all of those different departments and processes are connected through Bi-links. That means the customer's order can be fulfilled quickly and efficiently, without any unnecessary delays or complications.

But how do you get started with QBD? Well, the first step is to identify all of the different blocks in your business. That means looking at your departments, processes, systems, and even your people. Once you've identified all of those different blocks, the next step is to start connecting them through Bi-links. That means finding the connections between different parts of your business and building those connections up until you have a strong and cohesive web.

And the best part? QBD is modular and scalable, which means you can add or remove blocks as needed. That means you can adapt to changing circumstances and keep your business running smoothly no matter what.

So, there you have it folks: QBD in a nutshell. It's all about fitting the pieces of your business together in the right way and creating a strong and cohesive whole. And with Bi-links, your business will be more efficient, more effective, and more profitable than ever before. So, let's get started and see what QBD can do for you!

- Highlight the benefits of using QBD for your business.

Now, let's dive into the benefits of using QBD for your business.

First and foremost, QBD is all about transparency. With its decentralized ledger, all parties involved in a business process can see exactly what's going on at every step of the way. No more opaque supply chains or questionable business practices. QBD allows you to build trust with your customers and partners by providing complete transparency into your operations.

Secondly, QBD is incredibly secure. Because it uses cryptography and digital signatures, it's virtually impossible to tamper with the data stored on the QBD network. This means that you can rest easy knowing that your sensitive business information is safe from hackers and other bad actors.

Thirdly, QBD is highly efficient. By automating business processes through the use of smart contracts, you can reduce the need for intermediaries and speed up transaction times. This means you can get more done in less time, freeing up resources for other important tasks.

Fourthly, QBD is scalable. Whether you're a small business or a multinational corporation, QBD can grow and adapt to your needs. You can add or remove processes and components as needed, allowing you to tailor the platform to your specific business requirements.

Fifthly, QBD is cost-effective. By eliminating intermediaries and automating processes, you can save money on transaction fees and other expenses. Plus, because it's a decentralized platform, there's no need for costly middlemen or third-party service providers.

Sixthly, QBD is environmentally friendly. By reducing the need for paper-based processes and minimizing waste, QBD can help your business reduce its carbon footprint and contribute to a more sustainable future. As a matter of fact, some of our Distributed Ledger Technology is carbon-negative, meaning you may likely improve your ESG scores by adopting QBD.

Finally, QBD enables new business models. By allowing for secure and transparent data sharing and collaboration among multiple entities, QBD can open up new revenue streams and business opportunities. This can help you stay ahead of the competition and drive innovation in your industry.

So, there you have it, folks - the many benefits of using QBD for your business. By providing transparency, security, efficiency, scalability, cost savings, sustainability, and new business opportunities, QBD is truly a game-changer in the world of business.

- Encourage participants to ask questions and provide examples of how QBD could be applied in their specific roles.

Now, I'd like to open up the floor for questions and discussion.

First, I want to encourage everyone to ask questions. There's no such thing as a silly question, and I guarantee that if you have a question, someone else in this room has the same one. So please, don't hesitate to speak up and ask.

To help get the ball rolling, let me give you a few examples of how QBD could be applied in different roles within our company:

- *For our sales team, QBD could be used to securely and efficiently track and verify orders, as well as manage customer data and preferences.*
- *For our marketing team, QBD could be used to collect and analyze customer data, allowing us to better understand our target audience and personalize our messaging.*
- *For our finance team, QBD could be used to streamline payment processing and reduce the need for intermediaries, ultimately saving time and money.*

So, who has a question or example they'd like to share?

(Allow time for questions and discussion)

Here is a list of the top questions you can expect and solid answers you can rely on.

1. What is QBD?

Answer: QBD stands for Quantum Business Dynamics™ and is a methodology that enables businesses to build interconnected networks of data and value using blockchain technology.

2. How is QBD different from traditional business methods?

Answer: QBD is different from traditional methods in that it prioritizes data ownership and consent, transparency, and decentralized decision-making. It also uses blockchain technology to create secure, trustless networks of data and value.

3. How can QBD benefit my business?

Answer: QBD can benefit your business by increasing efficiency, transparency, and security in your operations. It can also help you build better relationships with customers and partners by prioritizing consent and data ownership.

4. How can I implement QBD in my business?

Answer: To implement QBD in your business, you should start by identifying key business processes that could benefit from increased efficiency and transparency. You should also work to educate your employees on the principles and benefits of QBD and invest in the necessary technology and training.

5. How does QBD handle data ownership and consent?

Answer: QBD prioritizes data ownership and consent by allowing individuals to own the rights to their data and exercise their consent elections through nonprofit information rights agents known as Authorized Agents.

6. How does QBD ensure security in its networks?

Answer: QBD uses blockchain technology to create secure, trustless networks that are resistant to hacking and other security threats. It also includes access controls and permissions that can be customized to meet the needs of individual businesses.

7. How does QBD handle multiple entities involved in a business process?

Answer: QBD uses smart contracts to automate business processes and reduce the need for intermediaries. It also enables data sharing and collaboration among multiple entities through its interconnected network of data and value.

8. How can QBD improve supply chain management?

Answer: QBD can improve supply chain management by enabling real-time tracking and verification of goods and services, providing valuable insights for optimization and decision-making, and creating secure and transparent supply chain networks.

9. How can QBD be used in the financial sector?

Answer: QBD can be used in the financial sector to create secure and efficient payment systems, reduce the need for intermediaries, and enable real-time monitoring and analysis of business processes.

10. How does QBD handle governance and accountability?

Answer: QBD includes a comprehensive system of governance and accountability that ensures all stakeholders are held responsible for their actions and decisions. This includes access controls, permissions, and consensus mechanisms.

11. How does QBD handle data validation?

Answer: QBD uses a variety of tools and techniques for data validation, including cryptography, digital signatures, and consensus mechanisms.

12. How does QBD handle scalability?

Answer: QBD is designed to be modular and scalable, allowing businesses to add or remove processes and components as needed to adapt to changing circumstances.

13. How does QBD handle interoperability with existing systems?

Answer: QBD is designed to be interoperable with existing systems and technologies, allowing businesses to integrate QBD seamlessly into their existing workflows and infrastructure.

14. How does QBD handle privacy concerns?

Answer: QBD prioritizes privacy by allowing individuals to own the rights to their data and exercise their consent elections through nonprofit information rights agents known as Authorized Agents. It also includes access controls and permissions that can be customized to meet the needs of individual businesses.

15. How does QBD handle regulatory compliance?

Answer: QBD includes a comprehensive system of governance and accountability that ensures regulatory compliance. It also includes access controls and permissions that can be customized to meet the needs of individual businesses.

16. Can QBD be integrated with existing systems and technologies?

Answer: Yes, QBD is designed to be interoperable with existing systems and technologies. This means that businesses can integrate QBD seamlessly into their existing workflows and infrastructure, without the need for major changes or disruptions.

17. How does QBD handle data ownership and rights management?

Answer: QBD handles the issue of data ownership and rights management through the use of Bridges, which are managed by nonprofit information rights agents called _RRMs. The _RRMs work as fiduciaries on behalf of the owners of the data, helping them to exercise their consent elections and ensuring that their data rights are protected.

18. How can QBD be used to create secure payment systems?

Answer: QBD can be used to create secure and efficient payment systems by leveraging the blockchain's cryptographic features to ensure that transactions are secure and transparent. This can reduce the need for intermediaries and minimize transaction costs, making payments faster and more cost-effective for businesses.

19. How can QBD be used in supply chain management?

Answer: QBD can be used in supply chain management to create a secure and transparent network. By using the QBD methodology, businesses can track goods and services in real-time, enabling them to verify the authenticity of the products and services they receive. This can help to reduce the risk of counterfeiting and fraud, while also improving the efficiency of the supply chain.

20. Can QBD be used in healthcare?

Answer: Yes, QBD can be used in healthcare to securely manage patient data, share information between healthcare providers, and track the delivery of healthcare services. By using QBD, healthcare providers can ensure that patient data is protected from unauthorized access and that information is shared only with authorized parties. This can help to improve the quality of care provided to patients while also reducing administrative costs and improving overall efficiency.

21. Is QBD suitable for small businesses?

Answer: Yes, QBD is suitable for businesses of all sizes. In fact, the modular and scalable nature of the QBD framework makes it particularly well-suited for small businesses that are looking to grow and expand over time. By starting with a basic QBD implementation and gradually adding additional processes and components, small businesses can easily adapt to changing circumstances and scale their operations as needed.

22. How can businesses ensure that their data is secure when using QBD?

Answer: Businesses can ensure that their data is secure when using QBD by implementing robust access controls and permissions, using encryption and digital signatures to protect data, and leveraging consensus mechanisms to ensure that all parties agree on the status of transactions. Additionally, businesses can choose to work with certified Agents and _RRMs who have been trained and certified by the Guild to ensure that their data is managed and protected in accordance with industry best practices.

23. How does QBD handle energy consumption?

Answer: QBD uses a proof-of-stake consensus mechanism, which requires significantly less energy than the traditional proof-of-work mechanism used by many other blockchain platforms. Additionally, QBD is designed to be modular and scalable, allowing businesses to customize the platform to fit their specific needs and requirements. This means that businesses can choose to use only the components and processes that are necessary for their operations, which can help to further reduce energy consumption. Ultimately, the energy consumption of QBD will depend on the specific use case and the size of the network, but the platform is designed to be energy-efficient and scalable.

Great questions, everyone! Now, for those of you who might have additional questions or concerns, I want to remind you that we have a wealth of resources available to us. We have a QBD expert on staff who can provide more information, as well as online forums and documentation that can be accessed at any time.

Additionally, if you have a specific use case in mind for QBD and you're not sure how to implement it, we can schedule a one-on-one session with a QBD consultant to help you brainstorm and strategize.

So, don't be shy about asking for help or seeking out more information. That's what we're here for.

Thank you all for your participation and engagement. I hope you're as excited about the possibilities of QBD as I am.

Session 2: Understanding the QBD Methodology In this session, we will delve deeper into the QBD methodology.

- Explain the six particles of QBD: Sources, Facts, Verifiers, Bridges, Binary Links, and Governors.

Suggestions:

Use a group exercise where participants are divided into four groups, with each group representing one of the particles. Each group would then have to work together to come up with examples of how their particle fits into the QBD framework and how it interacts with the other particles. This exercise would not only help participants understand the four particles, but also encourage collaboration and teamwork.

Another activity could be a role-playing game where participants act out different scenarios that involve the four particles. For example, one scenario could involve a company trying to secure their data using QBD. Participants would have to take on different roles, such as an Authority or a Verifier, and work together to create a secure and transparent system using the QBD framework.

Ultimately, the best game or activity to teach the four particles of QBD will depend on the specific audience and their learning style. It's important to choose activities that are engaging, interactive, and relevant to the participants' roles and responsibilities.

- Break down each particle and explain its importance.

Teacher: Alright, let's dive into the first particle of QBD: Sources. Sources are where it all starts, the origin of data. Sources can be anything that validates data, such as a sensor, a machine, or even a person. Without Sources, there would be no data to work with in the QBD system.

So, what is a Source? Well, in the world of QBD, a Source is a digital representation of a physical object, event, or entity that has been validated. It's like a virtual version of something in the real world. Sources are the building blocks of the QBD system, and they're the foundation upon which everything else is built.

Think about it like this: in the real world, we have physical objects like cars, houses, and people. In QBD, we have digital representations of those objects called Sources. And just like in the real world, these Sources have unique identities and attributes that make them distinct.

Now, you might be wondering, "Why are Sources so important?" Well, for one, Sources are what allow us to track and trace things in the digital world. They provide a way for us to know that something exists and to keep tabs on it as it moves through the system.

But the real power of Sources comes from their ability to carry information. Each Source can contain data about its physical counterpart, such as its location, condition, or history. This data can be updated and verified by other particles in the system, which ensures that the information is accurate and trustworthy.

For example, let's say you have a Source that represents a package being shipped from one location to another. That Source could contain information about the contents of the package, its destination, and its estimated delivery time. As the package moves through the shipping process, this information can be updated and verified by different particles in the system, such as Bridges and Verifiers. This ensures that everyone who needs to know the status of the package has access to accurate and up-to-date information.

Now, let's do a little puzzle to test your understanding of Sources. Imagine you have a Source that represents a car. What kind of information do you think that Source could contain? Take a moment to think about it, and then we'll discuss it as a group.

Great job, everyone! I'm hearing some great ideas, such as the car's make and model, its year, and its VIN number. All of those are excellent examples of the kind of information a Source could contain.

So, to summarize, Sources are the digital validation representations of physical objects, events, or entities in the QBD system. They allow us to track and trace things in the digital world and carry important information about those things. And most importantly, they're the foundation upon which everything else in the QBD system is built.

Next up, we'll be diving into a very important topic: Facts. Get ready for some more interactive fun!

Touchpoint 1: Now, let's have a little fun with this. Can anyone give me an example of a source of data in your business? Take a moment to think about it, and when you're ready, shout it out!

Student 1: Our cash registers! Student 2: Our website analytics! Student 3: Our social media channels!

Teacher: Great examples, everyone! Cash registers, website analytics, and social media channels are all excellent sources of data.

Sources in QBD are not just any data source. To be considered a QBD source, a data source must meet specific requirements. These requirements ensure that the data generated by the source is accurate, valid, and trustworthy.

Touchpoint 2: Here's a riddle for you all: What do you call data that you can't trust? Spoiler alert, the answer is: Spoofed data. It's not uncommon for sources to be compromised, and when this happens, it can be hard to tell if the data is real or fake. That's why it's crucial to ensure that Sources in QBD meet the requirements for accuracy, validity, and trustworthiness.

One way QBD ensures the accuracy of data from Sources is through the use of cryptography. Cryptography is a technique used to secure communication and protect sensitive information from being intercepted and modified. With QBD, cryptography is used to secure the data generated by Sources and ensure that it is not tampered with in transit.

Touchpoint 3: Alright, let's do one more interactive activity to solidify our understanding of Sources. In groups of three, I want you to brainstorm an idea for a new source of data that could be used in your business. It can be anything, but make sure it meets the requirements for accuracy, validity, and trustworthiness. You'll have five minutes to come up with your idea, and then we'll share them as a class.

By ensuring that Sources meet specific requirements for accuracy, validity, and trustworthiness, QBD is able to provide a system of data that is secure, transparent, and reliable. Without Sources, QBD would not be able to achieve its goal of creating a trusted fabric of data that can be used to automate business processes, enable real-time tracking and verification of goods and services, and more.

"Now that we have a good understanding of how Sources work, let's dive into another important particle: Targets. Imagine a game of darts, where the objective is to hit the bullseye. In QBD, Targets are like the bullseye. They are the relatable desire of a particular Binary Link.

A Target is the specific entity or data point that is being acted upon or affected by a particular Binary Link. For example, if the Source is a customer's order, the Target might be the successful delivery or the completed purchase process.

Targets play a crucial role in creating a Bi-link Fabric. They help us understand how different Sources are connected and what actions need to be taken to achieve the desired outcomes. Think of Targets as the endpoint of a short journey segment that we embark on when we create a Binary Link.

Now, let's get interactive again. I have a scenario for you. Imagine you work for an online clothing retailer, and you're trying to track inventory across your supply chain. You want to create a Binary Link between your warehouse and your online store to ensure that the inventory levels are accurate.

So, let me ask you, what would be the Target in this scenario?"

Student 1: "The Target would be the inventory levels, right?"

Teacher: "Absolutely, well done! The inventory levels are the specific data point that we want to act upon or affect. By creating a Binary Link between the warehouse and the online store, we can ensure that the inventory levels are accurate and up to date.

Now, let's look at another scenario. Imagine you work for a healthcare company, and you want to ensure that patient information is protected and secure. In this case, what would be the Target?"

Student 2: "The Target would be the patient information, right?"

Teacher: "Great job, exactly! The patient information is the specific data point that we want to protect and secure. By creating a Binary Link between the patient information and the security protocols, we can ensure that the data is safe and protected.

So, as you can see, Targets are a critical piece of the QBD puzzle. They help us understand the specific data points that we need to act upon and ensure that we're achieving our desired outcomes. With Targets, we can create a Bi-link Fabric that connects different Sources and helps us achieve our business goals.

Next, we'll move on to Verifiers and explore how they work in conjunction with Sources and Targets to ensure that our data is accurate and trustworthy. Stay tuned!"

Now that we've covered Sources and Targets, it's time to talk about the third particle in the QBD framework: Bridges.

Bridges serve as the connective tissue of the QBD network. They are the means by which Sources and Targets can interact with each other, and they allow for the creation of Binary Links that enable the exchange of verified data.

Think of a Bridge as a gatekeeper between different data sources. It's responsible for verifying the authenticity of the data before it can be added to the QBD network. This is an essential step, as it ensures that the data is reliable and trustworthy, which is critical when it comes to making informed business decisions.

Bridges are also responsible for managing access control and permissions within the QBD network. This means that they determine who can access which data and under what conditions. For example, if a company wants to share certain data with a partner, they can use a Bridge to create a Binary Link that only allows access to that data under specific circumstances, such as during a particular timeframe.

So, why are Bridges so crucial in the QBD framework? Well, they enable businesses to create a Bi-link Fabric that grows in value effortlessly over time. By connecting different data sources and creating Binary Links between them, businesses can create a seamless network that enables real-time data sharing and collaboration.

But, as with any other particle in the QBD framework, Bridges have their own set of challenges. One of the biggest issues is ensuring that the data being exchanged is accurate and reliable. To address this, QBD employs a system of verification that ensures that only validated data is allowed on the network.

Another challenge is managing access control and permissions. Since Bridges are responsible for controlling access to the data, it's essential to have robust and flexible access controls in place to ensure that only authorized parties can access the data they need.

Now, let's have a bit of fun with Bridges. Imagine you're playing a game of Bridge (the card game) with your friends. Each player is a different data source, and the cards represent the data being exchanged between them. The Bridge is the dealer, responsible for ensuring that each player gets the cards they need and that the game is played fairly. In this way, Bridges are the backbone of the QBD network, ensuring that data is shared securely and reliably.

So, that's Bridges in a nutshell. They're the connective tissue of the QBD network, enabling real-time data sharing and collaboration. And, with the right systems and controls in place, they can help businesses create a Bi-link Fabric that grows in value over time.

Alright, now let's move on to the final element of QBD - Binary Links. So, what exactly are Binary Links? Well, Binary Links are the digital tokens that are generated when a Source is successfully bridged to a Target within the QBD framework. These tokens represent the unique relationship and agreement between the parties involved in a business transaction.

Let's consider an example with two particles: Source A and Target B. Source A possesses actionable validation that is relevant to Target B, and they aim to establish an agreement. In the QBD process, a Bridge is created between Source A and Target B, signifying the brokered validation of their agreement. As a result of this Bridge, a Binary Link is generated. This Binary Link is a digital token that embodies the possessable value of the agreement, securely recording the details of the transaction.

By leveraging Binary Links, QBD ensures the immutability and transparency of the agreement record. Each Binary Link represents a unique relationship between transaction parties and serves as an unchangeable evidence of the agreement.

Now, let's engage in an interactive exercise to enhance your understanding of Binary Links. Imagine yourself as a Source, and the person sitting next to you as a Target. Your task is to create a Binary Link with that person, representing an agreement. Consider the specific terms and details you would include in the Binary Link to accurately capture the agreement.

[Pause for a minute]

Excellent! This exercise demonstrates how Binary Links function within the QBD framework. They are the digital tokens that result from the act of bridging a Source to a Target, encapsulating the unique relationship and agreement between the parties involved. These Binary Links provide a secure and unchangeable record of the transaction, ensuring trust and transparency.

To summarize, Binary Links are the digital tokens that are generated when a Source is successfully bridged to a Target in QBD. They represent the possessable value and unique relationship between transaction parties, providing a secure and unchangeable agreement record. This concludes our discussion on QBD particles and their relationships. I hope you've gained a better understanding of QBD and its potential benefits for your business. If you have any more questions or would like to explore QBD further, please feel free to reach out to me or any other member of the team."

- Highlight the relationships between the particles and how they work together to create a Bi-link Fabric.

An effective method for the teacher to teach about the relationships between the particles and how they work together to create a Bi-link Fabric is to use visual aids, such as diagrams or flowcharts, and to walk through examples step by step. This can help to make the abstract concept of the Bi-link Fabric more concrete and understandable.

Here is an example script for the teacher to use:

"Alright, let's delve into how all the particles work together to create the Bi-link Fabric. The Bi-link Fabric represents a network of interconnected relationships that facilitate the secure exchange of validated facts between various entities.

To illustrate this, let's consider two companies, Company A and Company B. Company A possesses Binary Links, which are derived from the data they collect about their customers. These Binary Links comprise Sources that encapsulate customer identity information, bridged to Targets that represent the permissions to share and, in specific cases, sell customer data like order information. Now, Company B wishes to access this data in order to fulfill customer orders. To facilitate this data exchange, both Company A and Company B establish their own protocol and policy decision points. These components rely on Binary Links as part of their authorization flows, effectively establishing a secure communication channel between the two companies.

In order to ensure that the shared data complies with the necessary permissions, Company B employs an automated process to verify the prior validations of identity encapsulated in the original Sources. By doing so, Company B can leverage Company A's investment in validation processes, eliminating the need for them to replicate the same infrastructure. The Verifier examines whether prior validations exist and if they align consistently and accurately with Quantum Business Dynamics™ (QBD™) standards.

The verification step adds another immutable data point to the transaction, with the platform storing records of each verification. Should Company B ever face legal challenges regarding their compliance with regulations, the platform provides immutable nonrepudiation evidence, attesting that Company B diligently fulfilled their obligations.

Once the facts have been verified by examining the Binary Link, the underlying security technology enables communication between the two companies. With confidence in the validated data provided by Company A, Company B can proceed with their work. QBD acts as a parallel assurance that the data can be used in a regulatory compliant manner.

Lastly, the platform, under the governance of its Governor, oversees the entire process, ensuring the smooth operation of adding confidence and proper management of the Particles.

As you can observe, each particle plays a crucial role in constructing the Bi-link Fabric. The Source captures validated identity data, acting as the foundation for further processing. The Validator guarantees the accuracy of the data, instilling trust in the exchanged information. Finally, the Broker of the Bridge establishes the necessary agreements, ensuring the seamless functioning of the overall system."

- Encourage participants to ask questions and participate in group activities.

What are the six particles of QBD?

Answer: The four particles of QBD are Sources, Targets, Bridges, and Binary Links.

What is the role of a Source in QBD?

Answer: A Source A Source is validated information that serves as the root Particle and starts off a new bi-link chain in QBD.

How do Bridges work in QBD?

Answer: A Bridge binds one Origin to one Target and represents an agreement between two humans. Bridges carry with them the concept of agency and rights resource management (_RRM), both of which help to manage the relationship represented by the Bridge and its related data uses and permissions.

What is the purpose of Verifiers in QBD?

Answer: Uses existing facts discovered through Handles to satisfy a question or request regarding the status of relationships. There can be many Verifiers, both human and machine.

What is a Binary Link in QBD?

Answer: Binary Links are the digital tokens that are generated when a Source is successfully bridged to a Target in QBD. They represent the possessable value and unique relationship between transaction parties, providing a secure and unchangeable agreement record.

What is the role of a Target in QBD?

Answer: A Target is a relatable desires established by an Authority that individually represent some desired destination or result.

How do Binary Links work with Bridges in QBD?

Answer: Binary Links and Bridges work together to ensure that there is integrity and utility in the established relationship. Bridges represent the agreement and its agency and information rights, and Binary Links contain the possessable value from that agreement.

How are permissions managed in QBD?

Answer: Permissions are managed through the use of a Bridge's rights resource management (_RRM) and the owner's consent. Parties must have the necessary permissions to access and share data.

What is the role of a Governor in QBD?

Answer: Governors are responsible for managing the platform and the main treasury. They oversee the implementation and governance of QBD.

How does QBD ensure data privacy and security?

Answer: QBD uses cryptography, access controls, distributed ledger, the Consent Name Service (CNS) and permission-based sharing to ensure data privacy and security. Only parties with the necessary permissions can access and share data, and the data itself is encrypted to protect against unauthorized access.

Activity: Bi-Link Fabric Exercise Divide the participants into small groups and provide each group with a set of QBD particles. Ask them to work together to create a Bi-link Fabric, connecting each particle in a way that reflects a real-life business scenario. Encourage them to think creatively and to use the particles in ways that are specific to their business processes.

Session 3: Implementing QBD in Your Organization

In this session, we will discuss how to implement QBD in your organization.

- Provide examples of how other organizations have successfully implemented QBD.

Case Study: ACME is a fast-growing e-commerce retailer that specializes in high-end fashion accessories. As the company expands and handles an increasing volume of customer data, they begin to experience challenges in managing data ownership and rights management. They struggle with ensuring customers have clear control over their data, and they worry about the potential for data breaches or unauthorized access. They want to find a solution that allows them to provide better data protection for their customers while also ensuring efficient data management within their own operations.

Solution: ACME decides to implement QBD as a solution for their data management and ownership issues. With QBD, they are able to provide customers with greater control over their data by enabling them to easily manage their consent elections. ACME is able to demonstrate their commitment to data protection and transparency to their customers by showing them how their data is managed and used within the QBD system. They are also able to streamline their internal operations by having clear and consistent protocols for managing data ownership and rights.

Welcome back, everyone! Now that we've covered the basics of QBD and how it works, let's take a look at some real-world examples of how other organizations have successfully implemented QBD.

Our case study today is about a fast-growing e-commerce retailer called ACME that specializes in high-end fashion accessories. As ACME expands, they begin to experience challenges in managing data ownership and rights management. They worry about ensuring customers have clear control over their data and protecting against the potential for data breaches or unauthorized access.

To address these challenges, ACME decides to implement QBD as a solution for their data management and ownership issues. With QBD, they are able to provide customers with greater control over their data by enabling them to easily manage their consent elections. ACME can show their customers how their data is managed and used within the QBD system, demonstrating their commitment to data protection and transparency.

Not only does QBD help ACME ensure efficient data management and customer data protection, but it also streamlines their internal operations. ACME now has clear and consistent protocols for managing data ownership and rights, making it easier to navigate data-related issues and reducing the risk of data mishandling.

This is just one example of how QBD can be successfully implemented to address data management and ownership issues. With its modular and scalable design, QBD can be customized to fit the unique needs and challenges of any organization. Whether it's a fast-growing e-commerce retailer like ACME or a multinational corporation with complex data structures, QBD can provide a solution for data ownership and rights management that is secure, efficient, and transparent.

That concludes our section on examples of successful QBD implementation. If you have any questions or would like to share your own experiences with QBD, please don't hesitate to speak up.

- Explain the importance of identifying key business processes and how to map them out using QBD.

Let's talk about the importance of identifying key business processes and how to map them out using QBD.

First, let's define what we mean by key business processes. These are the critical activities that our organization performs on a regular basis to deliver products or services to our customers. They are the processes that if they go wrong, can cause significant delays, customer dissatisfaction, or even lost revenue.

It's important to identify these key processes because they are the ones that we need to focus on improving, automating, or streamlining to achieve operational excellence. And this is where QBD comes in.

QBD provides us with a powerful tool to map out our business processes and identify the critical touchpoints where data is generated, processed, or transferred. By doing this, we can understand the flow of data within our organization and identify areas where we can improve data quality, reduce errors, or streamline processes.

So, how do we map out our business processes using QBD? Here are the steps:

Step 1: Identify the key business processes that you want to map out. This can be done by talking to process owners, subject matter experts, or reviewing process documentation.

Step 2: Create a high-level process map that shows the major steps in the process and the inputs and outputs at each step. This can be done using a flowcharting tool or even a whiteboard.

Step 3: Identify the key data elements that are generated, processed, or transferred at each step of the process. This includes both structured data (e.g., customer information, order details) and unstructured data (e.g., emails, documents).

Step 4: Determine the quality and accuracy of the data at each step of the process. This can be done by reviewing data sources, talking to data owners, or even performing data profiling.

Step 5: Identify areas where data quality can be improved, errors can be reduced, or processes can be streamlined. This can be done by analyzing the process map and data quality assessments.

Step 6: Implement changes to improve data quality, reduce errors, or streamline processes. This can include changes to process steps, data sources, data validation rules, or even the introduction of new technology.

Now, let's talk about an innovative way to teach this process.

We're going to play a game called "Process Mapping Challenge." Here's how it works:

- Divide the class into small groups of 4-5 people.
- Give each group a process map for a different business process in your organization.
- Challenge each group to identify the critical touchpoints where data is generated, processed, or transferred.
- Have each group present their findings to the class and discuss areas where data quality can be improved, errors can be reduced, or processes can be streamlined.
- Award points to each group based on the quality of their analysis and the creativity of their solutions.

This game is a fun and engaging way to teach the importance of identifying key business processes and mapping them out using QBD. It allows participants to practice the skills they've learned in a safe, low-stakes environment while also encouraging teamwork and collaboration.

Mapping out key business processes using QBD is indeed an important step towards achieving operational excellence. It allows us to identify areas where we can improve data quality, reduce errors, or streamline processes. And by playing the "Process Mapping Challenge" game, we can make the learning process fun and engaging for everyone involved.

- Discuss the role of the _RRM and GCA3 (Global Consent Authorized Agency Association) in QBD and how they can help with implementation.

Alright class, we've talked a lot about QBD today and how it can help businesses improve their processes and create a more transparent and efficient system. One aspect of QBD that we haven't touched on yet is the role of the _RRM and GCA3 in helping with implementation. To better understand this, let's first look at an example of industries that self-regulate.

Have any of you heard of the National Football League, or the NFL? Of course you have, it's a massive industry that generates billions of dollars every year. But did you know that the NFL has its own self-regulating body called the NFLPA? That stands for the National Football League Players Association, and its purpose is to ensure that players are treated fairly and that the league is operating in a way that is safe for players.

Another example is the National Restaurant Association. This industry organization represents over one million restaurant businesses and is responsible for setting standards, providing training, and advocating for the industry on various issues.

Now, you might be wondering what this has to do with QBD. Well, in a way, the role of the _RRM and GCA3 in QBD is very similar to these industry organizations. The _RRM, or the nonprofit authorized agent represented in the bridge, acts as an information rights agent for the owner of the data generated from their actions on the applicable system. They work to help the owner exercise their consent elections and ensure that their data rights are protected.

Similarly, GCA3, or the Global Consent Authorized Agents Association, acts as a guild that certifies _RRM candidates. Guilds in QBD are similar to the self-regulating bodies in the industries we just talked about. They provide oversight, set standards, and ensure that those who are working within their industry are following best practices.

So, just as the NFLPA ensures that the NFL is operating fairly and safely for players, and the National Restaurant Association sets standards for the restaurant industry, the _RRM and GCA3 in QBD work to ensure that data rights are protected, and best practices are being followed.

Now, I want you all to take a few minutes to think about how this concept of self-regulating bodies applies to your own industries or areas of expertise. How could a guild or self-regulating organization help to set standards and ensure best practices are being followed? And how does this relate to the _RRM and GCA3 in QBD?

(Pause for a few minutes to allow students to reflect and discuss)

Teacher: Great, let's share your ideas.

(Allow for some examples, and coach them to a better understanding of agency as it applies to their own industry)

Teacher: Super job! Now, while _RRMs, which are governed by their Guild GCA3 are satisfying a very specific for of agency called Information Rights Authorized Agency, which is a role supported by laws and regulations such as CCPA in California, your examples were perfect examples of agency and the Guild that manages them in your own agency. Whoever plays the role of Authority for creating the Targets in your business, they will get to set the agency at the Target level by selecting from a specific NAICS Industry Sector. This helps narrow the choices of who the agent will be that is assigned to each Bridge you create to that Target as well as automatically sets what group will be the Guild that certifies those agents.

To summarize, the _RRM and their Guild, GCA3, in QBD oversee best practices for data rights and consent management. And your Target appointed agent and their Guild will act as the agent for all other management of each bridge. This is similar to how industries self-regulate through organizations like the NFLPA or the National Restaurant Association. By ensuring that best practices are being followed and standards are set, we can create a more efficient and transparent system that benefits everyone involved.

- Encourage participants to ask questions and share any concerns they may have about implementing QBD in their organization.

Alright, class, now it's time for an open discussion. I want to hear your thoughts and concerns about implementing QBD in your organization. But I'm going to turn the tables on you a little. This time I'm going to be the student, and you are the teacher that will answer my questions.

Use your acting skills and take on the persona of different students will you ask the following questions to the class and coach them to answer correctly:

Student 1: I'm not sure if QBD will fit into our existing systems. We already have a lot of software in place, and I'm not sure if we can integrate it all with QBD.

Expected Answer: That's a valid concern. It's important to note that QBD is designed to be modular and scalable, so it can integrate with your existing systems seamlessly. We'll work together to identify the key business processes and workflows that can benefit from QBD, and we'll make sure to prioritize those for integration.

Student 2: What about training? How can we ensure that everyone in the organization understands how to use QBD?

Expected Answer: Another great question. We'll provide thorough training sessions for all employees to ensure that everyone understands the basics of QBD and how to use it effectively. We'll also provide ongoing support and resources to help with any questions or issues that may arise.

Student 3: How will QBD impact our data privacy policies?

Expected Answer: QBD actually enhances data privacy by allowing individuals to have more control over their personal data. With QBD, individuals can choose what data to share, with whom, and for what purpose. And the _RRM, as a nonprofit authorized agent, will work as a fiduciary on their behalf, ensuring that their data rights are protected.

Student 4: What about the cost? Is QBD expensive to implement?

Expected Answer: Actually, QBD is designed to be cost-effective and efficient. By streamlining workflows and reducing the need for intermediaries, QBD can actually save your organization money in the long run. And because it's modular, you can add or remove components as needed to fit your budget.

Student 5: How long does it take to implement QBD?

Expected Answer: That depends on the size and complexity of your organization, as well as the specific workflows you choose to integrate with QBD. But with proper planning and execution, we can have QBD up and running in as little as a few weeks.

Student 6: How can we measure the success of QBD in our organization?

Expected Answer: That's an excellent question. We'll work together to establish key performance indicators (KPIs) for QBD implementation, and we'll measure success based on those KPIs. We'll also gather feedback from employees and stakeholders to ensure that QBD is meeting their needs and providing value.

Student 7: What about security? How can we ensure that our data is protected?

Expected Answer: Security is a top priority for QBD. It uses advanced cryptography and digital signatures to ensure that data is secure and tamper-proof. And because QBD is decentralized, there is no single point of failure or vulnerability. We'll work with your IT team to ensure that QBD is implemented securely and in compliance with all relevant regulations and standards.

So, as you can see, there are many considerations when it comes to implementing QBD in your organization. But with proper planning, training, and support, we can make the transition smooth and successful. Thank you for your questions and concerns, and please don't hesitate to reach out if you have any further questions or feedback.

Mid-Morning Break

Session 4: Best Practices for Using QBD In this session, we will share best practices for using QBD in your organization.

- Discuss the importance of ongoing training and education for your team.

Today, we're going to talk about the importance of ongoing training and education for your team, and how you can set up a methodology for sustainable self-teaching within your organization. With QBD, there are some facets that are particularly appropriate for this approach.

First, let's talk about why ongoing training and education are so important. In today's rapidly changing business landscape, it's critical that your team stays up to date with the latest developments in your industry, as well as new technologies and methodologies that can help your organization stay ahead of the curve. By investing in ongoing training and education, you can help ensure that your team is equipped with the skills and knowledge they need to succeed.

One effective methodology for sustainable self-teaching is the 70:20:10 approach. This methodology breaks down training and education into three categories: 70% on-the-job experience, 20% social learning, and 10% formal training. Let's take a closer look at each of these categories and how they can be applied to QBD.

The first category, on-the-job experience, involves learning through practical experience and application. With QBD, this might involve assigning team members to work on specific Bridges or projects and allowing them to learn by doing. This approach can be particularly effective when combined with regular check-ins and feedback from supervisors or mentors.

The second category, social learning, involves learning from others, whether through collaboration, mentoring, or coaching. With QBD, this might involve setting up regular team meetings or knowledge-sharing sessions, where team members can discuss their experiences and learn from one another. This can be a great way to foster a culture of learning and continuous improvement within your organization.

The third category, formal training, involves more structured, classroom-style learning. With QBD, this might involve setting up regular workshops or training sessions, either in-person or online, to cover specific aspects of the QBD methodology. This could include topics such as data ownership and rights management, or the role of the _RRM and GCA3 in QBD.

In order to make this methodology sustainable, it's important to provide ongoing support and resources to your team. This might include access to online training materials, coaching or mentoring programs, or regular check-ins and feedback from supervisors or mentors.

Another key element of sustainable self-teaching is flexibility. With the 70:20:10 approach, it's important to be adaptable and willing to change course as needed. This might involve adjusting the mix of on-the-job experience, social learning, and formal training to better suit your team's needs and learning styles.

Ultimately, the goal of ongoing training and education is to create a culture of learning and continuous improvement within your organization. By investing in your team's skills and knowledge, you can help ensure that your organization remains competitive and successful over the long term.

So, to summarize, the 70:20:10 approach is an effective methodology for sustainable self-teaching, and there are many facets of QBD that are appropriate for this approach. By combining on-the-job experience, social learning, and formal training, and by providing ongoing support and resources, you can create a culture of learning and continuous improvement within your organization.

- Share tips on how to optimize your QBD Bi-link Fabric.

1. Start with a clear understanding of your business processes: Before implementing a Bi-link Fabric, take the time to understand your organization's key business processes. This will help you identify the areas where a Bi-link Fabric can be most beneficial.

Alright everyone, before we jump into the world of Bi-link Fabrics, let's start with the basics. The first step in optimizing your Bi-link Fabric is to have a clear understanding of your business processes. Think of it like trying to put together a puzzle without knowing what the picture looks like - it's going to be a lot harder! So, take the time to understand how your business functions and where the opportunities lie for a Bi-link Fabric to make a real impact. Trust me, your future self will thank you for the effort.

2. Define clear data rights and consent policies: Clear data rights and consent policies are essential to the success of a Bi-link Fabric. Be sure to establish policies that align with your organization's values and goals and communicate these policies to all stakeholders.

Alright folks, now we're moving on to tip number 2 for optimizing your Bi-link Fabric. And that is to define clear data rights and consent policies.

Now, I know this might sound like a boring bureaucratic task, but it's actually super important. You need to make sure everyone involved in your Bi-link Fabric understands their rights and responsibilities when it comes to data.

For example, let's say you're a healthcare provider using a Bi-link Fabric to track patient data. You need to make sure everyone involved knows exactly who owns that data, who can access it, and how it can be used.

You might be thinking, "But what if we have different data policies for different departments or types of data?"

Great question! And that's where QBD comes in handy. With QBD, you can set up different Bridges with different policies depending on the type of data involved. So, for example, you could have one Bridge for patient data with very strict policies, and another Bridge for internal employee data with more relaxed policies. The key is to make sure all policies are clear, consistent, and communicated to everyone involved.

By establishing clear data rights and consent policies, you'll help build trust and accountability within your Bi-link Fabric, which will ultimately lead to a more successful and valuable system. So, take the time to get these policies right, and make sure everyone is on the same page. Trust me, it'll be worth it in the end.

3. Use secure and reliable technology: The success of a Bi-link Fabric depends on the reliability and security of the technology used. Be sure to choose a technology that meets your organization's needs and has a proven track record of reliability and security.

When it comes to technology, it's not just about choosing the newest or trendiest option. The reliability and security of the technology you choose can make or break the success of your Bi-link Fabric. Think about it like building a house - you wouldn't want to use cheap and unreliable materials, would you? So, when it comes to your Bi-link Fabric, choose technology that you can trust to keep your data safe and secure. It's like building a strong foundation for your organization's future success.

4. Ensure data integrity and accuracy: Data integrity and accuracy are essential to the success of a Bi-link Fabric. Implement measures to ensure that data is accurate and reliable, and regularly monitor data for inconsistencies or errors.

Alright class, I'm sure we can all agree that nobody likes a liar, especially when it comes to data! So, make sure you're keeping it real with your Bi-link Fabric by ensuring data integrity and accuracy. No fibbing allowed! And just like with a bad poker face, if there are any inconsistencies or errors, make sure to call them out and address them. Remember, honesty is the best policy in both life and data management!

5. Create a culture of transparency and accountability: A Bi-link Fabric works best in a culture of transparency and accountability. Nurture this culture by promoting open communication and providing regular training and education on the importance of transparency and accountability.

Remember, fostering a culture of transparency and accountability is an ongoing process, but the benefits to your organization and stakeholders are well worth the effort. Keep promoting open communication and providing regular training and education to keep the culture strong and thriving. With a culture of transparency and accountability, your Bi-link Fabric will be able to achieve its full potential and bring value to your organization for years to come.

6. Regularly review and refine your Bi-link Fabric: Regularly review and refine your Bi-link Fabric to ensure that it continues to meet your organization's needs and goals. This may involve making changes to your policies or technology or adjusting your business processes.

It's important to stay proactive when it comes to your Bi-link Fabric. Take the time to review and assess its effectiveness on a regular basis, and don't be afraid to make necessary changes. Keep an open mind and be willing to embrace new ideas and technologies that can further optimize your Bi-link Fabric. Remember, your Bi-link Fabric is a living organism that can evolve and grow alongside your organization. By staying committed to regular review and refinement, you can ensure that your Bi-link Fabric remains a valuable tool for years to come.

7. Leverage analytics to gain insights: Analytics can provide valuable insights into the performance of your Bi-link Fabric. Use analytics to gain insights into your organization's performance and identify areas for improvement.

Alright class, time for some detective work! Think of analytics as a magnifying glass - it helps us uncover the tiny details we might have missed otherwise. With analytics, we can get a clearer picture of how our Bi-link Fabric is performing and find ways to make it even better. It's like being a spy, gathering intelligence to improve our operations. So, let's put on our detective hats and use analytics to uncover the secrets of our Bi-link Fabric!

8. Promote collaboration and teamwork: Collaboration and teamwork are essential to the success of a Bi-link Fabric. Encourage collaboration and teamwork by promoting open communication and providing opportunities for team members to work together.

Imagine your Bi-link Fabric is like a giant puzzle. Each piece represents a different aspect of your business, and it takes everyone working together to put it all together. By promoting collaboration and teamwork, you can ensure that all the pieces fit together seamlessly, creating a strong and efficient Bi-link Fabric. Encourage your team members to share ideas, provide feedback, and work together towards a common goal. Together, you can achieve great things and take your business to the next level!

9. Build strong relationships with stakeholders: Strong relationships with stakeholders are essential to the success of a Bi-link Fabric. Build strong relationships with stakeholders by engaging with them regularly, listening to their concerns, and addressing their needs.

Are you ready for the final stretch? Building strong relationships with stakeholders is the key to a successful Bi-link Fabric. Get excited about the potential for collaboration and trust that comes with strong stakeholder relationships. By engaging with stakeholders regularly, listening to their concerns, and addressing their needs, you can build a network of trust that will strengthen your Bi-link Fabric and drive business success. Let's finish strong by putting this into practice and building the strong stakeholder relationships that will take your organization to new heights!

10. Stay up to date with industry developments: Stay up to date with industry developments related to Bi-link Fabrics and data rights and consent management. This will help you stay ahead of the curve and ensure that your organization is using the latest and most effective practices.

And finally, let's talk about staying ahead of the game. As we've discussed, a Bi-link Fabric is an innovative solution for managing data rights and consent. But technology is constantly evolving, and it's important to stay up to date with industry developments.

(Pause for a moment)

So, how can you stay up to date? Attend industry conferences and workshops, subscribe to relevant publications and blogs, and network with industry leaders. By staying informed, you can ensure that your organization is using the latest and most effective practices.

(Pause again)

And that's a wrap on optimizing your Bi-link Fabric!

- Explain how to track and measure the success of your QBD implementation.

I. Introduction

Explain the importance of tracking and measuring QBD

Welcome to this lesson on tracking and measuring the success of your QBD implementation. As you know, QBD is a powerful methodology that can help businesses improve efficiency, reduce costs, and enhance security and transparency. However, in order to fully realize these benefits, it is important to track and measure the success of your QBD implementation. In this lesson, we will discuss why tracking and measuring QBD is important and provide you with some proven methodologies for doing so.

Why is Tracking and Measuring QBD Important?

1. *To Ensure the Success of the Implementation:*

Tracking and measuring QBD is critical to ensuring the success of the implementation. Without proper tracking and measurement, it can be difficult to determine whether the implementation is achieving the desired outcomes. By tracking and measuring QBD, you can identify any areas where the implementation may be falling short and take corrective action to address these issues.

2. *To Identify Areas for Improvement:*

Tracking and measuring QBD can also help you identify areas for improvement. By analyzing the data generated by the implementation, you can identify areas where processes can be streamlined or made more efficient. This can lead to cost savings and other benefits.

3. *To Meet Regulatory Requirements:*

Many industries are subject to regulatory requirements related to data privacy, security, and other areas. By tracking and measuring QBD, you can demonstrate compliance with these requirements and avoid penalties and other negative consequences.

Proven Methodologies for Tracking and Measuring QBD:

1. *Key Performance Indicators (KPIs):*

KPIs are metrics used to track progress towards specific goals. In the context of QBD, KPIs can be used to track progress towards the implementation's objectives. For example, KPIs might include the number of successful validations, the percentage of transactions completed without errors, or the time it takes to complete a transaction.

2. *Data Analytics:*

Data analytics involves analyzing the data generated by the QBD implementation to identify patterns, trends, and other insights. This can help you identify areas for improvement and optimize processes to improve efficiency and reduce costs. Data analytics can also be used to identify potential security or compliance issues before they become major problems.

3. *User Feedback:*

User feedback is an important tool for tracking and measuring the success of your QBD implementation. By soliciting feedback from users, you can identify areas where the implementation may be falling short and take corrective action. User feedback can also be used to identify areas where the implementation is working well and replicate these successes in other areas of the business.

So as you can see, tracking and measuring the success of your QBD implementation is critical to achieving the desired outcomes. By using proven methodologies such as KPIs, data analytics, and user feedback, you can identify areas for improvement, optimize processes, and demonstrate compliance with regulatory requirements. We hope that this lesson has provided you with a better understanding of the importance of tracking and measuring QBD, and some practical tools for doing so.

- Briefly introduce the concept of Key Performance Indicators (KPIs)

When implementing QBD, it's important to have a way to measure the success of your efforts. This is where Key Performance Indicators, or KPIs, come in. Think of KPIs as the dashboard of your QBD implementation - they help you monitor progress and identify areas for improvement.

But KPIs don't have to be dry and boring! Let's make them engaging. Imagine that you're driving a car on a road trip, and your KPIs are the dashboard gauges that help you keep track of your progress. Your speedometer tells you how fast you're going, your fuel gauge tells you how much gas you have left, and your GPS tells you how far you are from your destination. Without these gauges, you'd be driving blind - and that's no fun!

Similarly, KPIs help you navigate the road to success with QBD. They provide a way to measure the health of your system, identify areas for improvement, and ensure that you're on track to achieve your goals. And just like a road trip, implementing QBD can be an exciting journey. With the right KPIs in place, you'll be able to track your progress and celebrate your successes along the way.

So, let's embrace KPIs as the helpful dashboard gauges that they are. With the right KPIs in place, you'll be able to steer your QBD implementation in the right direction and enjoy the ride.

II. Identifying KPIs for QBD

- Define what KPIs are and how they work

So, what are KPIs? KPI stands for Key Performance Indicator, and just like the name suggests, they are key indicators that help you measure the performance of your business or organization. Think of them as the dashboard of your car, but for your business!

Now, KPIs work by providing you with a set of metrics that allow you to track your progress towards achieving your goals. These metrics can be anything from sales revenue, customer satisfaction, or even website traffic. By monitoring these KPIs, you can identify areas of improvement and adjust your strategy accordingly.

But wait, there's more! KPIs not only help you track your progress, but they also help you set goals and objectives. By setting specific, measurable, achievable, relevant, and time-bound (SMART) goals, you can ensure that your business is moving in the right direction.

And here's the fun part: imagine you're a superhero, and your KPIs are your superpowers. Just like a superhero uses their powers to fight crime, you can use your KPIs to defeat your business challenges and achieve your goals! So, let's suit up and get ready to conquer the business world with our KPIs!

Identify key areas of QBD to measure, such as data accuracy, process efficiency, and user satisfaction

First and foremost, data accuracy is crucial to the success of any QBD implementation. To measure data accuracy, we can track the percentage of validated particles and Binary Links, as well as the frequency of validation failures. This will give us a good idea of the overall accuracy of our data and help us identify any areas that need improvement.

Next, we need to measure process efficiency. This can be done by tracking the time it takes to complete key business processes, such as order fulfillment, payment processing, and supply chain management. We can also measure the number of errors or exceptions that occur during these processes, as well as the frequency of manual interventions. By measuring process efficiency, we can identify areas where we can streamline workflows, reduce costs, and improve overall performance.

Another key area to measure is user satisfaction. This can be done by conducting surveys or interviews with QBD users, to gather feedback on their experience with the system. We can also track metrics such as user adoption rate, user retention rate, and user engagement rate. By measuring user satisfaction, we can identify areas where we can improve the user experience, increase engagement, and drive adoption.

Additionally, we may want to measure the impact of QBD on key business metrics, such as revenue growth, cost savings, and customer satisfaction. For example, we can track the percentage of orders that are processed without error, the percentage of on-time deliveries, and the average order processing time. By measuring these metrics, we can see the direct impact of QBD on our business performance and make data-driven decisions to optimize our processes and improve our bottom line.

Finally, it's important to consider industry specific KPIs that may be relevant to your business. For example, a healthcare organization may want to track patient outcomes, while an energy company may want to measure carbon emissions reduction. By identifying and tracking industry specific KPIs, we can ensure that our QBD implementation is aligned with our overall business goals and objectives.

In summary, measuring the success of a QBD implementation requires a comprehensive approach that includes tracking key areas such as data accuracy, process efficiency, user satisfaction, and industry-specific metrics. By using KPIs to measure these areas, we can identify areas for improvement, optimize our processes, and drive business success.

- Brainstorm and prioritize specific KPIs for each area

Alright, everyone! Now that we've covered the importance of tracking and measuring QBD and learned about Key Performance Indicators, let's put our knowledge to the test with a fun interactive activity.

For this activity, we're going to brainstorm and prioritize specific KPIs for each area of QBD that we identified earlier - data accuracy, process efficiency, and user satisfaction. Here's how it's going to work:

1. Split into small groups of 3-4 people.
2. Each group will choose one area of QBD to focus on - data accuracy, process efficiency, or user satisfaction.
3. Using what you know about QBD and industry standard approaches on identifying KPIs, brainstorm as many specific KPIs as you can think of for your chosen area. Remember, KPIs should be measurable, specific, and relevant to the area you're focusing on.
4. Once you have a list of KPIs, prioritize them by importance. Which KPIs are most critical to the success of your QBD implementation? Which ones can you afford to track later on?
5. When you're finished, each group will share their list of KPIs and their prioritization with the class. We'll discuss the similarities and differences between the groups' KPIs, and we'll talk about the reasoning behind the prioritization.

To give you an idea of what you might come up with, here are some example KPIs for each area:

Data Accuracy:

- *Percentage of Bi-links validated by external sources*
- *Percentage of data entries that require correction*
- *Time taken to correct inaccurate data*

Process Efficiency:

- *Time taken to complete a transaction*
- *Number of transactions processed per hour/day/week*
- *Percentage of transactions completed without error*

User Satisfaction:

- *Average response time to user inquiries*
- *Number of user complaints*
- *Percentage of users who would recommend QBD to a colleague*

Remember, these are just examples - your group may come up with different KPIs that are more specific or relevant to your business. The key is to think about what metrics will give you the most valuable insights into the success of your QBD implementation, and to prioritize them accordingly.

Are you ready? Let's get started and see what KPIs we can come up with for our QBD implementation!

III. Setting Targets and Benchmarks

- Define what targets and benchmarks are and why they are important

The next step in measuring the success of your QBD implementation is to set KPI targets and benchmarks (Yes, we can us QBD in measuring our KPI – can you think of some good examples as we go through this next section?). These are important because they provide specific goals for your KPIs and help you track progress over time.

KPI targets are the specific values or outcomes that you want to achieve for each KPI. For example, if your KPI is process efficiency, your target might be to reduce the time it takes to complete a specific business process by 20%. Targets should be specific, measurable, and achievable.

Benchmarks, on the other hand, are the standard or reference point against which you measure your progress. Benchmarks can be internal, such as comparing current performance to past performance, or external, such as comparing your performance to industry averages or best practices.

Setting KPI targets and benchmarks can help you identify areas for improvement and measure progress towards specific goals. It can also help you identify areas where you are performing well and celebrate your successes.

What's exciting about QBD is that it can be used to capture and measure these targets and benchmarks within its own framework. Since QBD tracks data and metadata at every step of a business process, it can be used to set up Particles specifically for KPI targets and benchmarks. This means that not only can you measure the success of your main QBD fabric, but you can also use QBD itself to capture and measure the KPIs that are important to your business.

In this way, QBD becomes a powerful tool for continuous improvement, allowing you to identify areas for improvement, set specific goals, and track progress towards those goals within a secure and transparent framework.

- Set specific targets and benchmarks for each KPI identified in the previous section

In this section, we're going to talk about setting specific targets and benchmarks for each of those KPIs.

KPI targets and benchmarks are important because they provide a clear and measurable goal for each KPI. They help you understand where you are now, where you want to be, and how you're going to get there. Targets are specific, measurable, and time-bound goals that you set for each KPI. Benchmarks, on the other hand, are industry standards or best practices that you can use to compare your performance against others in your industry.

Let's take the example of a healthcare organization that has implemented QBD to improve their patient outcomes. One of the KPIs they identified is patient satisfaction, which they will measure using patient surveys. They might set a target of a 90% satisfaction rate within the next year. This target is specific (90% satisfaction rate), measurable (using patient surveys), and time-bound (within the next year).

To establish a benchmark for patient satisfaction, they might look at industry standards for patient satisfaction rates in similar healthcare organizations. If the industry standard is 80%, they can use this as a benchmark to compare their performance against others in their industry.

Another example could be a manufacturing company that has implemented QBD to improve their supply chain efficiency. One of the KPIs they identified is inventory turnover rate, which measures how quickly they're selling their inventory. They might set a target of increasing their inventory turnover rate by 20% within the next six months. This target is specific (20% increase in inventory turnover rate), measurable (using inventory turnover rate calculations), and time-bound (within the next six months).

To establish a benchmark for inventory turnover rate, they might look at industry standards for inventory turnover rates in similar manufacturing companies. If the industry standard is 10%, they can use this as a benchmark to compare their performance against others in their industry.

As you can see, setting targets and benchmarks for each KPI can help you track your progress and measure the success of your QBD implementation. It can also help you identify areas where you may need to make adjustments to your processes to achieve your goals.

When setting targets and benchmarks, it's important to make sure they're realistic and achievable. It's also important to communicate these targets and benchmarks to your team and stakeholders, so everyone is on the same page and working towards the same goals.

In conclusion, setting targets and benchmarks is an important step in measuring the success of your QBD implementation. By identifying specific targets and benchmarks for each KPI, you can track your progress, measure your success, and identify areas for improvement. Remember to make your targets and benchmarks realistic and achievable and communicate them clearly to your team and stakeholders.

- Discuss how to adjust targets and benchmarks over time as the QBD implementation progresses

Congratulations! Now that you have set your targets and benchmarks for your QBD implementation, let's talk about how to adjust them over time.

As with any project, it is important to continually monitor progress and adjust goals as needed. This is especially true with QBD, as it involves multiple stakeholders and processes. So, what are some things you can do to adjust your targets and benchmarks?

Firstly, you need to regularly review your KPIs and assess your progress. This can be done through regular reporting and analysis of the data collected through your QBD system. You should identify any areas where you are falling short of your targets and try to understand why this is happening. Are there specific processes or stakeholders that are causing delays or errors? Are there external factors that are impacting your progress?

Once you have identified the root cause of any issues, you can begin to adjust. This might involve changing the way certain processes are done, providing additional training to stakeholders, or investing in new technology or tools. The key is to be flexible and willing to make changes as needed to ensure that your QBD implementation is successful.

Let's consider a couple of examples. Imagine that your company has set a target for data accuracy, with a benchmark of 95%. However, after several months of implementing QBD, you find that your actual data accuracy is only 90%. Upon further investigation, you discover that a particular stakeholder is consistently making errors when entering data into the system. In this case, you might decide to provide additional training to this stakeholder to help them better understand how to use the QBD system.

Now, let's consider another example. Imagine that your company has set a target for process efficiency, with a benchmark of reducing processing times by 50%. However, after a few months of implementing QBD, you find that your actual processing times have only been reduced by 30%. Upon further investigation, you discover that the external system you are using to integrate with your QBD system is causing delays. In this case, you might decide to invest in a new integration tool or work with the external system provider to improve their integration capabilities.

As you can see from these examples, adjusting targets and benchmarks requires a combination of data analysis, problem-solving, and a willingness to invest in changes that will improve your QBD implementation. It is important to regularly review progress and be willing to adjust to ensure that you are on track to achieve your overall business goals.

So, we must remember, QBD implementation is an ongoing process that requires constant monitoring, analysis, and adjustment. By regularly reviewing your KPIs, assessing progress, and making necessary changes, you can ensure that your QBD implementation is successful and achieves the desired outcomes for your business.

IV. Tracking and Analyzing KPIs

- Discuss various tools and methods for tracking KPIs, such as dashboards and reports

Alright class, now that we have identified our KPIs and set our targets and benchmarks, it's time to talk about tracking our progress. There are a variety of tools and methods available to us for tracking our KPIs, such as dashboards and reports.

One useful tool that we can utilize is the QBD API. With this API, we can automate the creation of Particles for our KPIs, making it easier to track and measure our progress. This automation can help improve the accuracy and efficiency of our data collection, saving us time and effort.

In addition to the QBD API, there are many other tools and methods that we can use to track our KPIs. One such tool is the dashboard. A dashboard is a graphical representation of our KPIs, providing a quick and easy way to see our progress. Dashboards can be customized to display the specific KPIs that are important to us, and can be updated in real-time as new data becomes available.

Another useful tool is the report. Reports provide a more detailed analysis of our KPIs, allowing us to dive deeper into the data and identify trends and patterns. Reports can be generated on a regular basis, such as daily, weekly, or monthly, providing us with a consistent and reliable way to track our progress over time.

There are many other tools and methods available for tracking KPIs, and the key is to find the ones that work best for our specific needs and requirements. We can experiment with different tools and methods, and adjust them as needed over time as our QBD implementation progresses.

Let me give you a couple of examples of how this might work for our business. If we're a manufacturing company, we might set a KPI for production efficiency. We could use the QBD API to automatically create a Particle for each unit produced, and track the time it takes to produce each unit. We could then use a dashboard to display our production efficiency in real-time, and generate weekly reports to identify areas for improvement.

If we're a healthcare provider, we might set a KPI for patient satisfaction. We could use the QBD API to create a Particle for each patient visit, and track their satisfaction using a standardized survey. We could then use a dashboard to display our patient satisfaction scores over time, and generate monthly reports to identify trends and areas for improvement.

These are just a couple of examples of how we might use tools and methods to track our KPIs, but there are countless other possibilities. The key is to find the tools and methods that work best for our specific needs and requirements, and adjust them as needed over time to ensure that we are tracking our progress effectively.

- Highlight the importance of regular analysis and review of KPI data.

As you know, KPIs are an essential part of measuring the success of your QBD implementation. However, it's not enough to just set targets and benchmarks and track progress towards them. Regular analysis and review of KPI data is crucial to ensuring that your QBD implementation is running smoothly and efficiently.

So, why is regular analysis and review important? For one, it allows you to catch any issues or discrepancies early on before they become larger problems. By regularly monitoring your KPI data, you can identify areas that are not meeting your targets or benchmarks and take action to address them. Additionally, regular analysis and review can help you identify trends and patterns in your data, which can inform future decision-making and help you make more informed business decisions.

Let's say, for example, that one of your KPIs is data accuracy. You've set a target for 99% accuracy, and you've been tracking your progress towards this goal for the past few months. After regular analysis and review of your KPI data, you notice that your accuracy has actually decreased slightly over the past month. Without regular analysis and review, you may not have caught this decrease in accuracy until it had become a larger issue. However, because you caught it early, you're able to investigate and address the issue, preventing it from becoming a larger problem.

So, how do you conduct regular analysis and review of KPI data? There are a variety of tools and methods available, such as dashboards and reports. Dashboards can provide a quick snapshot of your KPI data, allowing you to easily see how you're tracking towards your targets and benchmarks. Reports, on the other hand, can provide more in-depth analysis of your KPI data, allowing you to dive deeper into trends and patterns.

One tool that you may find particularly useful is the QBD API. As you know, QBD is designed to be platform-agnostic, which means that it can be used with a variety of existing systems and technologies. The QBD API allows you to automate the creation of Particles, making it easier to track and measure your KPIs. By leveraging the QBD API, you can ensure that your KPI data is accurate and up to date, without the need for manual data entry.

In addition to tools and methods, it's important to establish a regular cadence for analysis and review. Depending on the nature of your business and your KPIs, this could be weekly, monthly, or quarterly. Whatever the cadence, it's important to stick to it and ensure that all stakeholders are aware of when analysis and review will occur.

Let's keep in mind that regular analysis and review of KPI data is critical to ensuring the success of your QBD implementation. By catching issues early and identifying trends and patterns, you can make more informed business decisions and ensure that your QBD implementation is running smoothly and efficiently. So, make sure to establish a regular cadence for analysis and review, leverage available tools such as dashboards and reports, and consider using the QBD API to automate Particle creation.

- Provide examples of how KPI data can be used to identify areas for improvement and make data-driven decisions

Let's take a look at how the data you collect can be used to make data-driven decisions and identify areas for improvement.

By regularly analyzing and reviewing your KPI data, you can identify trends and patterns in your business processes and operations. This can help you make informed decisions about how to improve your QBD implementation and achieve your goals.

Let me give you a few examples of how KPI data can be used to identify areas for improvement:

First, let's say you're tracking the data accuracy KPI. You notice that there are several instances where the data entered into the system is incomplete or incorrect. This can be a sign that there is a problem with your data entry process or that employees need more training on how to enter data correctly. By identifying this issue through your KPI data, you can take action to address the root cause and improve data accuracy moving forward.

Another example is if you're tracking the process efficiency KPI. You notice that certain steps in your business process are taking longer than they should, causing delays and inefficiencies. This could be an indication that your process needs to be optimized or that certain steps can be automated using QBD's smart contract capabilities. By identifying these bottlenecks through your KPI data, you can make informed decisions about how to improve process efficiency and streamline operations.

Finally, let's say you're tracking the user satisfaction KPI. You notice that there is a decrease in user satisfaction ratings for a particular process or feature of your QBD implementation. This could be an indication that the process or feature is not meeting the needs of your users or that there is a problem with the user experience. By identifying this issue through your KPI data, you can gather feedback from your users and take action to improve their experience, ultimately increasing overall user satisfaction.

As you can see, KPI data can be a powerful tool for identifying areas for improvement and making data-driven decisions. By regularly analyzing and reviewing your KPI data, you can make sure that your QBD implementation is on track and continuously improving over time.

But how can you make sure you're getting the most out of your KPI data? One way is to use available tools, such as dashboards and reports, to visualize your data and make it easier to analyze. Additionally, QBD's API can be used to automate Particle creation, enabling you to collect and analyze data more efficiently and effectively.

Remember, the key to success with KPIs is to regularly analyze and review your data and take action to address any issues or areas for improvement. By doing so, you can ensure that your QBD implementation is achieving its goals and continuously improving over time.

V. Continuous Improvement

- Emphasize the importance of continuous improvement in QBD implementation

Class, we've covered a lot about setting targets and benchmarks, identifying and measuring KPIs, and analyzing the data to make informed decisions. But there's one more thing that's crucial for the success of your QBD implementation: continuous improvement.

Continuous improvement means always striving to do better and looking for ways to make your QBD implementation more efficient, effective, and secure. And this is important because the business landscape is constantly evolving, and you need to adapt to stay ahead of the competition.

One way to continuously improve your QBD implementation is by regularly reviewing your KPI data and identifying areas for improvement. For example, if you notice that your data accuracy KPI is consistently below the target, you can investigate why this is happening and take corrective action.

Another way to continuously improve is by soliciting feedback from your users, such as employees, customers, and partners. You can use surveys or focus groups to gather feedback on what's working well and what needs improvement. And you can use this feedback to make data-driven decisions on how to optimize your QBD implementation.

Finally, you can leverage the power of QBD itself to continuously improve. QBD is designed to be flexible and adaptable, allowing you to create custom workflows and business rules to fit your specific needs and requirements. And you can use QBD's data validation and smart contract features to automate business processes and reduce the need for intermediaries.

In conclusion, continuous improvement is critical for the success of your QBD implementation. By regularly reviewing your KPI data, soliciting feedback from users, and leveraging QBD's flexibility and automation features, you can stay ahead of the competition and achieve your business goals. So, let's keep striving to do better and make our QBD implementation the best it can be.

- Discuss how KPI data can be used to identify areas for improvement and prioritize changes

As we discussed earlier, tracking and measuring KPIs in your QBD implementation is crucial to understanding how your business processes are performing and making data-driven decisions to improve them. But once you have that data, what do you do with it? How do you use it to make meaningful changes?

One of the most important things you can do is use KPI data to identify areas for improvement. For example, let's say that you are tracking the KPI of process efficiency and you notice that one particular process is consistently taking longer than it should. By analyzing the data, you may be able to pinpoint the exact step in the process that is causing the delay and take steps to address it.

In addition to identifying specific areas for improvement, KPI data can also be used to prioritize changes. You may find that there are several areas that need improvement, but not all of them are equally important. By analyzing the data and prioritizing the areas that will have the biggest impact on overall performance, you can make the most efficient use of your resources and see the greatest improvement in your QBD implementation.

One way to prioritize changes is to use a technique called the Pareto principle, also known as the 80/20 rule. This principle states that 80% of the effects come from 20% of the causes. Applied to QBD KPI data, this means that 80% of your performance issues are likely caused by 20% of your processes or factors. By identifying and addressing that critical 20%, you can see significant improvements in your overall performance.

Another way to prioritize changes is to use a technique called impact analysis. This involves analyzing the potential impact of each change on your QBD implementation and prioritizing them based on the expected return on investment. For example, if you have identified a process that is causing delays and you are considering investing in a new technology to address it, you would want to weigh the potential benefits of that technology against the cost of implementing it to determine if it is a worthwhile investment.

Ultimately, the key to using KPI data to identify areas for improvement and prioritize changes is to use it as a guide, not a mandate. The data can provide valuable insights into your QBD implementation, but it is up to you to interpret that data and make decisions based on it. By taking a thoughtful and data-driven approach to continuous improvement, you can maximize the benefits of your QBD implementation and stay ahead of the competition.

- Encourage ongoing review and adjustment of KPIs and targets to ensure that they remain relevant and effective

Now that we have covered the importance of continuous improvement, let's discuss the ongoing review and adjustment of KPIs and targets. As with any system, QBD implementation is not a one-time set-and-forget solution. It requires ongoing attention and refinement to ensure that it remains relevant and effective for your business.

Regular review and adjustment of KPIs and targets are necessary to ensure that they are still relevant and aligned with your business goals. As your business evolves and changes, so should your KPIs and targets. This will help you to stay focused on what matters most and to make data-

driven decisions that drive continuous improvement.

Using KPI data, you can identify areas that require improvement and prioritize changes based on their impact on your business goals. For example, if your data accuracy KPI shows that you have a high error rate in your data inputs, you may want to prioritize improvements in data collection and validation processes. Or, if your user satisfaction KPI shows that your customers are not happy with the usability of your QBD implementation, you may want to prioritize improvements to the user interface and experience.

To ensure that your KPIs targets and benchmarks remain effective, you should also regularly review and adjust them. This may involve adding new KPIs or removing outdated ones, adjusting targets based on your current performance, or changing the metrics you use to measure KPIs to better align with your business goals.

For example, if your original target for data accuracy was to maintain a 95% accuracy rate, but your current performance shows that you consistently exceed that target, you may want to adjust your target to a higher percentage to continue driving improvement. Or, if you find that a particular KPI is no longer relevant or does not align with your business goals, you may want to remove it and replace it with a more relevant KPI.

Ultimately, the ongoing review and adjustment of KPIs and targets is critical to ensuring the ongoing success of your QBD implementation. By regularly analyzing KPI data and making data-driven decisions, you can continue to improve your business processes, reduce costs, increase efficiency, and drive better outcomes for your customers and stakeholders.

VI. Conclusion

- Summarize the importance of KPIs in measuring QBD success

In summary, KPIs are critical in measuring the success of a QBD implementation. By identifying and tracking specific KPIs, businesses can gain valuable insights into the accuracy, efficiency, and satisfaction of their QBD processes. With the use of targets and benchmarks, they can set specific goals and monitor progress towards achieving them. Regular analysis and review of KPI data can help identify areas for improvement and inform data-driven decisions. By utilizing available tools and methods, businesses can automate the creation of particles and track KPIs through dashboards and reports. It is important to remember that KPIs and targets should be regularly reviewed and adjusted to remain relevant and effective. Continuous improvement is crucial for achieving long-term success with QBD implementation.

- Encourage ongoing tracking and analysis of KPI data for continuous improvement.

As we come to the end of our discussion on KPIs in QBD implementation, I want to emphasize the importance of ongoing tracking and analysis of KPI data for continuous improvement. By regularly monitoring and reviewing KPI data, you can identify areas for improvement and make data-driven decisions that will help your QBD implementation succeed.

It's important to remember that KPIs are not set in stone and should be reviewed and adjusted over time to ensure they remain relevant and effective. As you implement QBD, you may discover new KPIs that need to be tracked or existing KPIs that are no longer relevant. By regularly reviewing and adjusting your KPIs and targets, you can ensure that they continue to provide valuable insights and help drive your QBD implementation forward.

In addition, KPI data can be used to identify areas for improvement and prioritize changes. For example, if you discover that data accuracy is a major issue, you can prioritize efforts to improve data validation and verification processes. By using KPI data to inform your decisions, you can ensure that your efforts are focused on the areas that will have the greatest impact on QBD success.

Finally, I want to encourage you to make continuous improvement a key part of your QBD implementation strategy. As you track and analyze KPI data, you may uncover new opportunities for improvement or discover that certain approaches are not working as well as you had hoped. By embracing a culture of continuous improvement and being willing to adapt and make changes, you can ensure that your QBD implementation continues to evolve and improve over time.

In summary, KPIs are a crucial tool for measuring QBD success and driving continuous improvement. By tracking and analyzing KPI data, you can identify areas for improvement, prioritize changes, and ensure that your QBD implementation is on track to achieve your goals. So, I encourage you to continue to use KPIs as a key part of your QBD implementation strategy and to make ongoing tracking and analysis of

KPI data a top priority.

- Encourage participants to share their own experiences and insights on how to use QBD effectively.

To wrap up our discussion on measuring QBD success, I want to encourage everyone to share their own experiences and insights on how to use QBD effectively. While we've covered a lot of ground today, there is always more to learn, and we can all benefit from hearing each other's perspectives.

Remember, QBD is a powerful tool for creating decentralized, secure, and transparent business processes. By tracking and measuring KPIs, we can gain valuable insights into how our QBD implementation is performing and identify areas for improvement.

But KPIs are not a one-size-fits-all solution. Every business is unique, and the KPIs that are important for one organization may not be as critical for another. That's why it's important to regularly review and adjust your KPIs to ensure that they remain relevant and effective.

Additionally, we talked about the importance of ongoing analysis and review of KPI data. By regularly analyzing KPI data, we can identify areas for improvement and make data-driven decisions to optimize our QBD implementation.

Finally, I want to emphasize the importance of continuous improvement in QBD implementation. By continually striving to improve our business processes, we can create more efficient, effective, and secure workflows that benefit everyone involved.

So, I encourage everyone to continue tracking and analyzing their KPI data, regularly reviewing and adjusting their KPIs, and seeking out new ways to use QBD to improve their business processes. Thank you all for participating in this discussion, and I look forward to hearing more about your experiences with QBD in the future.

Activity: QBD Implementation Plan Divide the participants into small groups and ask them to work together to create an implementation plan for QBD in their organization. Provide them with a template and encourage them to think about how QBD can be applied to their specific business processes.

Session 5: Conclusion and Next Steps

In this final session, we will wrap up the training and discuss next steps.

- Recap what has been covered in the training session.
- Discuss any outstanding questions or concerns.
- Provide participants with resources for ongoing learning and support.
- Encourage participants to share their feedback on the training session.

Thank you for attending this onboarding session for QBD. We hope that you have found this training valuable and informative. Remember that QBD is an ongoing process of learning and improvement, and we encourage you to continue to explore its potential for your organization.

RESOURCES

Glossary of Terms

Agent: A person or entity authorized to act on behalf of another person or entity. In QBD, agents manage Bridges for the Owner.

Authority: An organization or entity with the power to make decisions and enforce rules. In QBD, authorities create the Target and set its attributes.

Binary Link: A type of particle in QBD that represents a bi-directional relationship between a Source and a Target.

Bridge: A type of particle in QBD that represents a relationship between a Source and a Target, where a Broker establishes the relationship.

Broker: A person or entity that facilitates transactions between two or more parties. In QBD, brokers build Bridges as a result of an agreement between two people.

CID: Content Identifier, a unique hash value that identifies data in a distributed ledger.

Confidential Feature: A type of attribute in QBD that contains sensitive data and is not publicly available.

Data Partner: A company that collaborates with another company to share data or information.

Governor: The person who manages the whole QBD platform and owns and manages the main treasury. There is one Governor per platform. The Governor is responsible for appropriately managing the invoicing of charges for underlying services, such as blockchain gas prices, to various parties based on the rules of QBD as set in Particles by the people of appropriate roles such as Authorities when they create Targets.

Guild: A self-regulating organization that certifies Agents and _RRMs. In QBD, the guild that manages the _RRMs is GCA3. Each industry has their own approach to what we call Guilds. Most of the time, industries are self-regulated, but in some cases, they are regulated.

Handle: A unique identifier for a particle that is used to retrieve and validate the particle's state.

Holder: The person who currently possesses a token (Source or Binary Link) in their wallet.

Owner: A person or entity that has legal rights and control over a particular asset or property. In QBD, the owner is the person who owns the information rights for the Source and who can provide direction to the _RRM for consent management regarding their related information rights.

Public Feature Particle: A type of particle in QBD that contains publicly available attributes about a particle.

_RRM: Rights Resource Manager. In QBD, an _RRM is a nonprofit authorized agent represented in the bridge, responsible for acting as an information rights agent for the owner of the data generated from their actions on the applicable system. They work to help the owner exercise their consent elections and ensure that their data rights are protected. The _RRM manages the information rights particular to the industry represented by a bridge's agency for the Owner to ensure their consent elections are honored.

Serial number: A publicly available identifier that indicates the order of creation of a particle in a series of particles.

Series number: A publicly available identifier that indicates the Class or Series of a particle on a distributed ledger.

Source: A type of particle in QBD that represents an Origin or starting point in a relationship.

Target: A type of particle in QBD that represents an end point or destination in a relationship.

Validate: To confirm the accuracy, truth, or validity of something through a process of proving that it meets a set of criteria or standards. Validation does not necessarily provide permanent confirmation of accuracy, as new information or evidence may arise that undermines or contradicts the original validation at which point it would be considered invalidated.

Validator: A person or entity responsible for verifying the authenticity and accuracy of data or information. In QBD, validators create Sources as a result of validating facts. For Sources, they validate claims regarding some distinctive of identity, and for Bridges, they are validating facts about an agreement between two people.

Verify: To check and confirm the accuracy or validity of something based on a prior validation from a trusted source. Verification is a process of checking the state of validation to confirm that it is still accurate and has not changed to invalid. Verification is a single moment in time, and once something has been verified, it is considered to be accurate or valid at that moment, with the understanding that it may need to be re-verified in the future if circumstances change or new information emerges. Unlike the word validate, there is no action of "inverify".

Verifier: A person or entity responsible for checking and confirming the validity of data or information. In QBD, verifiers use existing facts discovered through handles to satisfy a question or request regarding the status of relationships. There can be many verifiers, both human and machine.

Veracity: A hashed feature of a particle in QBD that is used to ensure the authenticity and integrity of the particle's data.

References and Further Reading

McKinsey & Company. (n.d.). Digital Transformation: The $100 Trillion Prize - A special collection of insights to help companies seize opportunities at the intersection of technology and business. Retrieved from https://www.mckinsey.com/business-functions/organization/our-insights/unlocking-success-in-digital-transformations

Deloitte. (2020). 2020 Manufacturing Industry Outlook. Deloitte Insights. Retrieved from https://www2.deloitte.com/us/en/insights/industry/manufacturing/manufacturing-industry-outlook.html

Accenture. (2017). The Cost of Compliance. Accenture. Retrieved from https://www.accenture.com/us-en/insights/financial-services/cost-of-compliance

CBC News. (2020). Pharma company fined $5.1 million for privacy breach involving patient data. Retrieved from https://www.cbc.ca/news/business/privacy-breach-fine-1.5702216

Global News. (2021). Ontario cannabis producer hit with $300K fine for violations. Retrieved from https://globalnews.ca/news/7716703/ontario-cannabis-producer-fined-violations/

The Guardian. (2020). Westpac to pay record $1.3bn fine over money laundering and child exploitation breaches. Retrieved from https://www.theguardian.com/australia-news/2020/sep/24/westpac-to-pay-record-13bn-fine-over-money-laundering-and-child-exploitation-breaches

World Tourism Organization. (2020). COVID-19 Related Travel Restrictions: A Global Review for Tourism. Retrieved from https://www.unwto.org/international-tourism-and-covid-19

McKinsey & Company. (2014). Cybersecurity: An essential business priority. Retrieved from https://www.mckinsey.com/business-functions/risk/our-insights/cybersecurity-an-essential-business-priority

U.S. Department of Health and Human Services. (2020, June 1). OCR imposes a $2.3 million civil money penalty for HIPAA violations. Retrieved from https://www.hhs.gov/hipaa/for-professionals/compliance-enforcement/agreements/CHSP/index.html

CNBC. (2020, October 28). Major US retail chain hit by apparent cyberattack, exposing customer data. Retrieved from https://www.cnbc.com/2020/10/28/major-us-retail-chain-hit-by-apparent-cyberattack-exposing-customer-data.html

Financial Conduct Authority. (2019, December 19). FCA fines bank £44m for failings in treatment of customers in arrears. Retrieved from https://www.fca.org.uk/news/press-releases/fca-fines-bank-44m-failings-treatment-customers-arrears

Ellen MacArthur Foundation. (2017). A new textiles economy: Redesigning fashion's future. Retrieved from https://www.ellenmacarthurfoundation.org/publications/a-new-textiles-economy-redesigning-fashions-future

Nielsen. (2018). The Sustainability Imperative: New Insights on Consumer Expectations. Retrieved from https://www.nielsen.com/us/en/insights/report/2018/the-sustainability-imperative/

Cone Communications. (2016). 2016 Cone Communications Millennial Employee Engagement Study. Retrieved from https://www.conecomm.com/research-blog/2016-millennial-employee-engagement-study.

Gartner. (2018). Market Guide for Data Preparation. Retrieved from https://www.gartner.com/doc/3881366/market-guide-data-preparation

McKinsey & Company. (2012). The social economy: Unlocking value and productivity through social technologies. Retrieved from https://www.mckinsey.com/business-functions/organization/our-insights/the-social-economy

Retail Dive. (2018, October 5). Study: Retail executives see collaboration as key to innovation. Retrieved from https://www.retaildive.com/news/study-retail-executives-see-collaboration-as-key-to-innovation/539908/

Harvard Business Review. (2017, November 13). The High Cost of Poor Communication. Retrieved from https://hbr.org/2017/11/the-high-cost-of-poor-communication.

Deloitte. (2018). Cross-functional teams: The key to break down silos and drive innovation. Retrieved from https://www2.deloitte.com/content/dam/Deloitte/de/Documents/Innovation/Deloitte-Innovation-Cross-Functional-Teams.pdf

KPMG. (2021). The top 10 business risks facing companies in 2021. Retrieved from https://home.kpmg/xx/en/home/insights/2021/01/top-10-business-risks-facing-companies-in-2021.html

Deloitte. (2020). Navigating uncertainty: 10 business risks to watch. Retrieved from https://www2.deloitte.com/content/dam/Deloitte/us/Documents/risk/us-risk-navigating-uncertainty-10-business-risks-to-watch.pdf

PwC. (2021). Top 10 risks in 2021: Resilience during uncertainty. Retrieved from https://www.pwc.com/us/en/services/consulting/risk-regulatory/top-risks.html

HealthLeaders Media. (n.d.). IT-Clinical Collaboration Key to Unlocking Telehealth's Full Potential. Retrieved from https://www.healthleadersmedia.com/clinical-care/it-clinical-collaboration-key-unlocking-telehealths-full-potential

Retail Dive. (n.d.). The cost of a lack of collaboration. Retrieved from https://www.retaildive.com/news/the-cost-of-a-lack-of-collaboration/593678/

Harvard Business Review. (2016). Collaborative overload. Retrieved from https://hbr.org/2016/07/collaborative-overload

Deloitte. (n.d.). The future of manufacturing: Deloitte University Press. Retrieved from https://www2.deloitte.com/content/dam/Deloitte/us/Documents/technology-media-telecommunications/us-tmt-pred16-manufacturing.pdf

Simpson, P. (2019). The business case for sustainability: a literature review. Journal of Cleaner Production, 208, 1551-1565.

Business Roundtable. (2019). Statement on the Purpose of a Corporation. Retrieved from https://www.businessroundtable.org/business-roundtable-redefines-the-purpose-of

ABOUT THE AUTHOR
and Contributors

J. Oliver Glasgow is a visionary leader in the field of identity and privacy solutions with a proven track record of success. With extensive experience in the telecom, banking, commerce, and insurance industries, he has shaped product vision, developed and executed strategies, assembled and led teams, and managed budgets to bring real-world applications to life. In addition to his professional achievements, Jay is a talented writer and composer, creating original stories and music that reflect our times. He is also a dedicated philanthropist, using his skills as a supervisor for Habitat for Humanity to make a positive impact in his community. With his expertise in identity, privacy, and trust relationships, Jay has helped countless clients achieve their goals.

Alan Nekhom is an experienced and innovative technology executive with a strong aptitude for identifying client needs and bridging technology gaps. He has created and implemented profitable products and services across a wide range of industries and is a leader in process development, technology delivery, and implementation. With a track record of success spanning over two decades, he has consistently demonstrated his ability to influence future products, technologies, and business strategies, ultimately empowering both internal and external clients with innovative solutions. Alan's expertise lies in cloud transformation, architecture, and the evaluation of emerging technologies to maximize revenue potential. He excels at building and leading strong teams that foster a sense of ownership and urgency while consistently delivering results that meet or exceed expectations.

Stuart Davis is a seasoned global executive with over 33 years of progressive leadership roles focusing on remediating, transforming, and delivering risk management and compliance programs. He is passionate about integrating talent and technology with compliance and regulatory risk management to achieve transformative outcomes to support both business and regulatory objectives. He has a proven track record of leading financial institutions through the AML and financial crimes regulatory environment. Stuart is a thought leader, speaker, and policy advocate on the social impact of financial crime such as Human Trafficking, Modern Slavery, Wildlife trafficking, and Online Child Exploitation. With his expertise in tokenization, distributed ledger technology, big data and cloud, robotics, AI, and machine learning as it relates to fighting financial crime and operational cost optimization, Stuart is at the forefront of innovation in his field.

www.ingramcontent.com/pod-product-compliance
Lightning Source LLC
LaVergne TN
LVHW022156200926
848481LV00006B/1765
9798218310301